The Great Pheromone Myth

Searching for snarks. Illustration by Henry Holiday from Lewis Carroll's *The Hunting of the Snark: An Agony in Eight Fits.*

The Great Pheromone Myth

RICHARD L. DOTY

The Johns Hopkins University Press

Baltimore

The Johns Hopkins University Press
2715 North Charles Street
Baltimore, Maryland 21218-4363
www.press.jhu.edu

Library of Congress Cataloging-in-Publication Data

Doty, Richard L.
The great pheromone myth / Richard L. Doty.
    p.   cm.
Includes bibliographical references and index.
ISBN-13: 978-0-8018-9347-6 (hardcover : alk. paper)
ISBN-10: 0-8018-9347-X (hardcover : alk. paper)
 1. Pheromones.   2. Mammals—Physiology.   I. Title.
QP572.P47D68 2009
573.9′29—dc22   2009009801

A catalog record for this book is available from the British Library.

Frontispiece: Searching for snarks. Illustration by Henry Holiday from Lewis Carroll's *The Hunting of the Snark: An Agony in Eight Fits.* Macmillan, New York, 1891.

Special discounts are available for bulk purchases of this book. For more information, please contact Special Sales at 410-516-6936 or specialsales@press.jhu.edu.

The Johns Hopkins University Press uses environmentally friendly book materials, including recycled text paper that is composed of at least 30 percent post-consumer waste, whenever possible. All of our book papers are acid-free, and our jackets and covers are printed on paper with recycled content.

In a speech delivered at the German League of Human Rights in 1932, Albert Einstein proclaimed that "the most beautiful and deepest experience a man can have is the sense of the mysterious. It is the underlying principle of religion as well as all serious endeavors in art and science." It is to those scientists seeking to understand the mystery of the relationship between organisms and their chemical environment that this book is dedicated. Without their unrelenting efforts, the important role of olfaction, gustation, and the other chemical senses in biology, ecology, and medicine would not be realized.

Contents

....

The pheromone concept, derived for insects, was first applied to mammals in the early 1960s. Although there is disagreement as to what constitutes a mammalian pheromone, such an agent is most commonly viewed as an innate biologically derived chemical that induces a well-defined behavioral or endocrine response in an invariant and species-specific manner. In nonhuman mammals, such diverse functions as sex recognition, courtship, copulation, fighting, nesting, social communication, maternal behavior, and altered endocrine function have been attributed to pheromones. In humans, the choice of a mate, the length of menstrual cycles of women living in close quarters, and even the selection of a seat in a dentist's waiting room have been said to be influenced by pheromones. Whether humans have pheromones was heralded by *Science* magazine in 2005 as one of the top 100 outstanding scientific questions of the era, emphasizing the importance of this issue to the scientific community at large.

In this book I provide a historical overview of the pheromone concept and the myriad attempts to apply it to mammals. Major studies of the effects of excretions and secretions on mammalian social behaviors and endocrine responses are reviewed in detail, and a critical evaluation is made of claims that such responses are due to pheromones. Based upon both empirical and theoretical grounds, I conclude that mammalian pheromones do not exist despite our continued fascination with the pheromone concept, numerous claims of the chemical isolation of pheromones, and the expenditure of millions of dollars on the part of industry and government to find such entities.

The reasons for my conclusion are multiple, as will become apparent throughout this volume. One problem with the pheromone concept is that it dichotomizes stimuli and complex behaviors into two classes—pheromonal and non-pheromonal—logically precluding the existence of multiple classes and continua. A second problem is the nominal fallacy, that is, the tendency to confuse naming with explaining. A third problem with this concept is that it assumes one or, at most, a few species-specific molecules of innate origin, largely impervious to learning and distinct from other types of chemical stimuli, are the motive influences. This is rarely the case in mammals. The interpretation of multiple molecules within specific contexts by dynamic and plastic nervous systems largely defines their biological effects. Learning, which in some cases occurs even before birth, plays a significant role in determining or modulating most odor-mediated mammalian behavioral responses, including a number that involve the accessory, that is, vomeronasal, olfactory system. Although a case can be made that exposure to urine and a number of its natural constituents may influence, perhaps inherently, the endocrine system of mice via the vomeronasal system, multiple chemicals are involved and species specificity is questionable. In some cases such stimuli perturb physiological homeostasis, altering reproductive function via stress. Rather than focusing on one or a few molecules, as per the pheromone concept, evolution seems to have favored adaptations related to the detection of multi-chemical features of the complex and changing chemical environment critical for survival.

I am aware that my perspective is heretical and may disturb the sensibilities of some scientists. I am also aware that it is difficult to prove the negative, particularly when definitions lack operational substance and become transmuted repeatedly to fit idiosyncratic contingencies. I only ask that the reader keeps an open mind as the material in this volume is traversed. Although I am under no illusion that this treatise will reverse long-standing beliefs that pheromones mediate many mammalian behaviors, it is gratifying that a growing number of neurobiologists are moving away from the simplistic tenets of the pheromone concept in both their research and theorizing. Hopefully this book will increase awareness, particularly among young scientists, of the complexity of chemical communication and the nature of the influences of biologically derived chemicals on a range of reproductive and other behaviors.

I thank the editorial staff of the Johns Hopkins University Press for their willingness to consider my radical perspective on this topic and the National Institutes of Health, whose financial support has largely made my career possible. I am grateful to those colleagues and friends who provided feedback or encouragement to me during the course of the project, including my intellectual mentor Jack King, Solange Chadda, Mary Lou Derksen, Carol Doty, Lee Drickamer, Daniel Hashimoto, Eva Heuberger, Robyn Hudson, Jane Hurst, Daqing Li, William Marra, Michael Meredith, Jessica Morton, Barbara Rolls, Burt Slotnick, James B. Snow, Jr., Isabelle Tourbier, Jon Treem, and Donald A. Wilson.

The Great Pheromone Myth

ONE

Introduction

....

The delusion is extraordinary by which we thus exalt language above nature—making language the expositor of nature, instead of making nature the expositor of language.

Alexander B. Johnson, *A Treatise on Language*, 1836

It has been nearly a half century since the term *pheromone* was first applied to chemically mediated mammalian behaviors and endocrine responses (Parkes and Bruce, 1961; Wilson and Bossert, 1963; Whitten, 1966) and 40 years since Alex Comfort published his speculative but influential *Nature* paper entitled "Likelihood of Human Pheromones" (Comfort, 1971). Where has the field of mammalian pheromone investigation gone since then? What progress has been made? Where are the mammalian pheromones that have been so actively sought? What, in fact, are pheromones?

Frank Beach, in his classic critique of contemporary comparative psychology entitled *The Snark Was a Boojum* (Beach, 1950), reminded us of Lewis Carroll's poem, *The Hunting of the Snark* (Carroll, 1874). In this story of "the impossible voyage of an improbable crew to find an inconceivable creature," a baker, a banker, a bellman, a beaver, and a menagerie of other disparate characters went on a trip to hunt snarks. But how was a snark to be recognized? The bellman pointed out that the snark habitually got up late, had a poor sense of humor, was arrogantly ambitious, and had a taste that was "meager but hollow." A number of species of snarks were about, some with whiskers who scratched and others with feathers who bit. When the baker heard that a few snarks were boojums, he fainted. After being revived, he recalled the words of his late uncle:

If your Snark be a Snark, that is right:
Fetch it home by all means—you may serve it with greens
And it's handy for striking a light...

But oh, beamish nephew, beware of the day,
If your Snark be a Boojum! For then
you will softly and suddenly vanish away,
And never be met with again!

It is the thesis of this book that quests to find and identify mammalian pheromones have largely been snark hunts and that pheromonology has become the modern phrenology of chemosensory science, providing simple, but false, explanations of nearly every type of chemically mediated social behavior and endocrine response imaginable. Like snarks, mammalian pheromones seem to come in all colors and flavors, but when captured frequently prove to be boojums, eluding chemical identification and, in rare instances, heralding the disappearance of the involved scientists from the scientific literature. Why is this so? Because the basic tenet of the pheromone concept—that one or a few hormone-like species-specific chemicals are the messages—is generally fallacious when applied to mammals. In this group of vertebrates, chemically mediated social processes are rarely hardwired, and most chemicals involved in their social communication are idiosyncratic and comprised of multiple compounds, a number of which are affected by stress, diet, and other factors. Even when neuroendocrine responses are involved, the influences come from multiple chemicals and depend as much on the receiver as the sender, commonly being influenced by the receiver's prior experience and motivational state. Clearly, the meaning of information transferred via the chemical medium does not reside solely in the stimulus.[1]

As will become apparent throughout this book, nearly all phenomena classically attributed in mammals to "releaser" pheromones, that is, chemical agents said to elicit stereotypic behavioral responses, depend upon learning, context, or novelty. Many of those attributed to "primer" pheromones, that is, ones that alter endocrine function, reflect physiological and psychological responses to abnormal changes in the social and physical environment. Some such changes reflect alterations in homeostasis brought on by stress. Without negating the fact that biological secretions ubiquitously influence mammalian social and sexual behaviors, as well as endocrine state, I will show why most such influences cannot be divorced from higher-order cognitive processes and rarely differ fundamentally from similar influences of other sensory systems. I will emphasize the fact that the main and accessory olfactory systems, which often work in tandem with one another, encode peripheral stimulus information that is collated and synthesized at some point by brain structures whose elements are honed by experience. Since, in many cases, different chemicals or sets of chemicals can lead to the same endpoint and the chemical message can reflect individuality, gender, diet, emotional state, and other factors, a definitive élan vital is illusive.

In light of these and other considerations, conceptualizing physiological and behavioral changes in mammals as responses to invariant, species-specific, and hormone-like chemicals is misleading. It is my view that such thinking unwittingly supports the stereotypic concept of chemosensory primitiveness, distorting and oversimplifying what is known about chemosensory physiology. Moreover, dividing stimuli into the dichotomous classes of pheromones and non-pheromones is illogical, as it precludes the existence of multidimensional and interactive classes or continua, as discussed in detail in Chapter 8. The term *pheromone,* with its inherent linguistic baggage and lack of operational tethering, would seem to be of no more scientific value for describing chemically mediated behaviors or endocrine responses than such hypothetical terms as *audiomones, visuomones,* or *touchamones* would be for describing analogous phenomena induced by nonchemical stimuli.

Some critics will view my argument as simply one of semantics. This is not the case. A key element of my thesis is that it is erroneous to infer that a plurality of mammalian behaviors and endocrine responses is uniquely determined in an invariant way by single or small sets of chemical stimuli and to apply a generic and misleading name to the presumptive agents in support of such an inference. My thesis is not, however, totally independent of semantics. As noted by Pinker (2007), "Semantics is about the relation of words to reality—the way that speakers [or writers] commit themselves to a shared understanding of the truth, and the way their thoughts are anchored to things and situations in the world" (p. 3). Clearly, the belief that simple and presumptive hormone-like chemical agents, signified by the term *pheromone,* are responsible for an extensive array of behaviors and endocrine states assumes a reality that is questionable, as will be shown in the chapters that follow.

This volume is composed of seven chapters in addition to the present one. Chapter 2 illustrates how surprisingly little unanimity exists among dictionaries, textbooks, and scientists as to what, in fact, are pheromones. In the case of mammals, some scientists assume such agents are volatile and work through the olfactory system, whereas others believe they are largely nonvolatile, are commonly odorless, and primarily operate via the vomeronasal system, a system that, as discussed in Chapter 7, is nonfunctional in humans. Others opine that mammalian pheromones only come from specific glands separate from urine or saliva. Many believe that they are unconsciously perceived, whereas others debate this point. In Chapters 3 and 4, the important and overlooked role of learning in establishing the meaning of chemicals involved in mammalian social communication is addressed, with an emphasis on scent marking in Chapter 4. These two chapters provide the reader with an overview of the complexity of chemically mediated mammalian behaviors and how the pheromone concept, if anything, hinders the explanation of such behaviors. Chapters 5 and 6 demonstrate, on a case-by-case basis, how major claims of the identification of releaser and primer pheromones either are not reproducible, are based on findings in which learning or novelty plays the primary role, or provide chemicals that fall short of truly or uniquely mimicking the actual behavioral or endocrine processes observed in the normal circumstance. In many cases, pheromonal involvement has been inferred without the identification of any such chemical or chemicals. Chapter 7 specifically addresses the question of human pheromones, demonstrating that claims of such agents are fraught with questionable inferences and, in most cases, are based upon poor research and statistically compromised findings. Chapter 8 seeks dialectic closure.

TWO

What Is a Mammalian Pheromone?

....

The difference between Newtonian physics, which was falsified by Einstein's theory of relativity, and astrology, lies in the following irony. Newtonian physics is scientific because it allowed us to falsify it, as we know that it is wrong, while astrology is not, because it does not offer conditions under which we could reject it. Astrology cannot be disproved, owing to the auxillary hypotheses that come into play. Such point lies at the basis of the demarcation between science and nonsense.

Nassim Taleb, *Fooled by Randomness*, 2004

The pheromone concept, when applied to mammals, seems to have morphed into multiple identities, making it difficult, like astrology, to tie down as a scientific entity. That being said, this concept—in its broadest sense—has profoundly and indelibly influenced the modern zeitgeist in which the nature of chemosensation, indeed biology itself, is viewed. This influence extends across all levels of biological science, from the molecular study of gene expression and regulation, sensory transduction, and neuroendocrine function to the macroscopic study of social behavior. As of this writing, on the order of 8,000 research papers have employed the term *pheromone,* with 90% focusing on invertebrates (Figure 2.1). Of the remainder, the majority have focused on mammals.

Pheromones have been reported in nearly all mammalian orders, save those lacking functional olfactory systems (e.g., whales, porpoises, and other members of Cetacea). Particularly within the Rodentia, which accounts for over forty percent of all named species of mammals, the world of chemicals associated with feeding, fighting, drinking, nesting, courtship, copulation, predator-prey relations, social communication, and endocrine

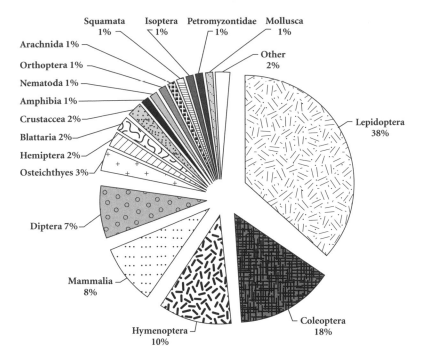

FIGURE 2.1. The proportion of published research efforts on pheromones divided on the basis of taxonomy. These proportions were derived from 7,802 counts of papers taken from the Web of Science (Thompson Scientific, USA, http://scientific.thompson.com/products/wos) using the search term "pheromone" and combinations of the scientific and common names for major taxonomic groups. The segment of the chart marked "others" includes, in decreasing order of research publications, Neuroptera, Thysanoptera, Trichoptera, Aves, Annelida, Collembola, Dermaptera, Cnidaria, Myriapoda, Thysanura, Onychophora, Odonata, Strepsiptera, Echinodermata, Platyhelminthes, Phasmida, and Crocodylla. Modified from Symonds and Elgar (2008).

function has been split into two mutually exclusive camps, the pheromonal and the non-pheromonal, with the majority falling into the pheromonal. E. O. Wilson, an entomologist and a pioneer of the field of sociobiology, succinctly provided an evolutionary argument for pheromones in 1970:

> Evolutionary inference, together with substantial new experimental evidence from studies in animal behavior and natural products chemistry, leads to the conclusion that chemical communication is the paramount

mode of communication in most groups of animals. In the early evolution of animal behavior, chemical releasers, or pheromones as they are now generally called, were probably also the first signals put to service. We know that communication among protozoan cells must have preceded the origin of the metazoans, and this primitive signaling was almost certainly chemical. Consider then the possibility that pheromones are in a special sense the lineal ancestors of hormones. (Wilson, 1970, p. 133)

Wilson provided a table showing the chemical structures of pheromones from various phyla, including the Chordata, and made it crystal clear that such agents were present within land-dwelling vertebrates: "Sex attractants, both male and female, are widespread in amphibious reptiles and mammals, although they are still poorly documented in both groups... These pheromones are now known to be common in primates, including even the female rhesus" (p. 135).

Wilson's suggestion that pheromones are lineal ancestors of hormones would seem in conflict with evidence that many insects employ chemicals found in their diet and food plants directly as pheromones. As noted by Hendry et al. (1975), "Diversification of insect species may be primarily due to the pheromone complexes available during evolution of host plants, rather than to the evolution of separate insect communication systems independent of changes in certain dietary chemicals" (p. 62). Wilson's perspective also contrasts with that of the ecologist Vero Wynne-Edwards (1962), one of the early proponents of group selection. While acknowledging the ubiquity of chemical communication throughout the animal kingdom, Wynne-Edwards never employed the term *pheromone* and espoused a somewhat less encompassing perspective on the evolution of chemicals involved in social communication:

Most functional odours have presumably been derived through natural selection from metabolites excreted or secreted in the first place for some different purpose. Integumental glands originally supplying mucus or wax to the skin have probably often been elaborated in this way to produce the relatively powerful odours commonly employed as signals, for recognition and other purposes. (p. 91)

Aside from the idea of Wilson and others in the 1960s and 1970s that pheromones were hormones or precursors to hormones was the idea that they somehow would lead to a much better understanding and quantification of the social behaviors of mammals. Gleason and Reynierse (1969) stated the following in an early comprehensive and optimistic review:

> As communication media, pheromones afford an interesting opportunity to investigate the development of a communication system based on a unit, the pheromone, permitting precise quantification and identification. The information transmitted by the components of a visual signal can never be quantified in the same manner as a specific dilution of a chemical distributed over a given area, even though the duration of the effect of that chemical cannot always be regulated. Within a species, interesting questions can be raised regarding the amount and type of information transmitted by a given pheromone in different dilutions or in combinations with different, more general, olfactory cues and/or visual or auditory signals. (p. 63)

Today, few scientists would argue that such optimism has been fulfilled. A key question underlying such lack of fulfillment is what, in fact, are pheromones?

Origin of the Pheromone Concept

In the 1930s entomologists sought to bring order to the myriad chemicals employed in insect communication and reproduction. As described by Stuart (1970), numerous approaches to classifying such agents were initially tried, including classification according to (1) chemical structure alone, (2) their source of origin within the animals that secreted or excreted the agents, and (3) the behavioral or endocrine responses that seemed to be released or induced by their actions. It was the latter classification scheme that eventually took hold. This was due to the chemicals' relatively direct ties to observable events, the technical problems associated with identifying the involved chemicals, and the sanctification of the instinct concept by ethologists. Some ethologists readily generalized such theoretical concepts as innate releasing mechanisms and fixed action patterns to all vertebrates, including humans. The latter focus on instinct was comforting to many biologists, as this meant that complex behaviors, including those of vertebrates, could more or less be tied directly to genetics in a straightforward fashion, being analogous to morphological markers useful in taxonomy. However, as pointed out by Daniel Lehrman in his classic paper entitled "A Critique of Konrad Lorenz's Theory of Instinctive Behavior" (Lehrman, 1953), labeling behaviors as instinctual—while perhaps gratifying—produces an either/or dichotomy with significant associated dangers. For example, such labeling tends to preclude the need to study developmental or experiential factors associated with the fruition of a given behavior, oversimplifying even its genetic underpinnings.

It will become apparent throughout this volume that the term *pheromone* suffers from many of the same limitations as the term *instinct* when mammals are involved. Indeed, a parallel can be drawn between the pheromone/non-pheromone dichotomy and the ancient instinct/anti-instinct controversy represented in more modern times by McDougall's instinct doctrine (McDougall, 1921) and the anti-instinct views of such notables as Dunlap (1919), Titchener (1928), and Yerkes (1911). Yerkes concluded that "instinct is one of those historical concepts which has been overgrown by meaning. It is so encrusted with traditional significance that it is almost impossible to use it for the exact descriptive purposes of science" (p. 378). Similarly, Titchener (1928) noted, "Instinct has long been one of those catchwords of popular psychology... and it does scientific harm... by its profession to explain, to name the cause of action" (p. 462).

The term *pheromone* formally came into existence to replace the term *ectohormone* (external hormone) in insects. In 1932 the entomologist Bethe distinguished between hormones secreted within the body (endohormones) and those excreted outside of the body (ectohormones), dividing the latter agents into those with intraspecific effects and those with interspecific effects (termed *homoiohormones* and *alloiohormones,* respectively) (Bethe, 1932).[1] Over two decades later, Karlson and Lüscher (1959) replaced the term *homoiohormone* with the term *pheromone,* defining pheromones as "substances which are secreted to the outside by an individual and received by a second individual of the same species, in which they release a specific reaction, for example, a definite behavior or a developmental process" (p. 55). However, "certain overlaps between closely related species may occur." Importantly, these authors distinguished between pheromones that act via olfaction and those that act via oral or ingestive routes, with the former producing immediate releasing responses (e.g., initiating and guiding the flight of the male silkworm moth, *Bombyx mori,* to the female) and the latter producing delayed endocrine or reproductive responses (e.g., the caste-determining and reproduction-inhibiting substances of many social insects).[2]

Although other schemata have since been proposed to overcome difficulties inherent in the initial classification system of insect excretions and secretions used in communication (e.g., Brown, 1985; Stuart, 1970), the term *pheromone* has remained the standard for describing chemically mediated *intraspecific* invertebrate interactions. This is in spite of the fact that, as pointed out by Hölldobler and Carlin (1987), "Most insect semiochemicals have proven to be complex mixtures, and single-compound pheromones are actually rare" (p. 567). Nonetheless, single chemical insect pheromones do exist, as exemplified by bombykol (E-10,Z-12-hexadecandien-1-ol), a sex attractant excreted by the female silkworm moth, *Bombyx mori* (Figure 2.2) (Butenandt et al., 1959). The identification of this agent was a triumph in modern chemistry, reflecting a 20-year-long effort to isolate 6 mg of active material from over a half million moths.

Generalization of the Pheromone Concept to Mammals

Whitten (1975) attributes the first demonstration of a mammalian releasing pheromone to Kelley (1937), who reported that vaginal secretions from an estrous ewe induced a ram to copulate with a pregnant ewe, an inappropriate mating partner. He credits Andervont (1944) for providing the first example of a mammalian primer pheromone (Whitten and Bronson, 1970). However, the first use of the term *pheromone* in describing mammalian behavior surfaced in the early 1960s (Wilson and Bossert, 1963; Whitten, 1966). In an influential review appearing in *Science,* Parkes and Bruce (1961) differentiated between "chemical messengers" that acted within the individual (e.g., hormones and "other excitatory substances" such as carbon dioxide) and pheromonal agents acting between individuals via ingestion, absorption, or sensory receptors, and concluded that "endocrinology has flowered magnificently in the last 40 years; exocrinology is now about to blossom" (p. 1054). The pheromone concept was further popularized by Wilson in a 1963 *Scientific American* article on the topic. Wilson explicitly set the tone for the nature of both priming and

FIGURE 2.2. The female silkworm moth *Bombyx mori*. The male detects E-10, Z-12-hexadecandien-1-ol, a chemical compound that serves a primary role in attracting the male to the female. Photograph courtesy of John Pelafigue (http://myphotos.ws).

releasing mammalian pheromones. In the case of putative mammalian releasing pheromones, he focused on musks and noted:

> Pheromones that produce a simple releaser effect—a *single specific response* mediated directly by the central nervous system—are widespread in the animal kingdom and serve a great many functions [italics mine]. The chemical structures of six attractants are shown [in the picture] on page 108. Although two of the six—the mammalian scents muskone and civetone— have been known for some 40 years and are generally assumed to serve a sexual function, their exact role has never been rigorously established by experiments with living animals. In fact, mammals seem to employ musk-like compounds, alone or in combination with other substances, to serve several functions: to mark territories, to assist in territorial defense and to identify the sexes. (Wilson, 1963, p. 101)

The caption of the picture in which the chemical structures of civetone and muskone were shown read as follows: "Six sex pheromones including the identified sex attractants of four insect species as well as two mammalian musks generally believed to be sex attractants. The molecular weight of most sex pheromones accounts for their narrow specificity and high potency."

That same year, Wilson and Bossert (1963) reiterated the distinction between "primer" and "releaser" effects. Releasing pheromones elicited specific behaviors and priming pheromones produced neuroendocrine or developmental changes. In their review, in which only an invertebrate example was employed, these two types of effects were deemed non-mutually exclusive: "It is quite possible for the same pheromone to be both a primer and a releaser; e.g., 9-ketodec-2-enoic acid, which inhibits ovary development, inhibits queen-cell building by workers and attracts males during the nuptial flight" (p. 675).

It was not until the late 1960s and early 1970s that claims of the chemical isolation of mammalian pheromones were made. The initial claims were for pheromones of the releasing type and included (1) a set of aliphatic acids isolated from the vaginal secretions of rhesus monkeys that was said to elicit copulatory behaviors in males (Michael and Keverne, 1968, 1970a, b; Curtis et al., 1971; Keverne and Michael, 1971; Michael et al., 1971); (2) agents within the tarsal scent glands of male black-tailed deer that elicited licking by females (Brownlee et al., 1969; Müller-Schwarze, 1971; Müller-Schwarze et al., 1974); (3) a substance from the midventral scent gland of Mongolian gerbils that received investigation from other gerbils (Thiessen et al., 1974); and (4) two steroids from the submaxillary salivary glands of boars that facilitated lordosis in female pigs (namely, 5α-androsten-16-en-3-one and its related alcohol) (Melrose et al., 1971). As will be described in detail in Chapter 5, responses to such agents are not invariant and are significantly influenced by learning and context, contrary to the tenor of the pheromone concept.

Definitions and Redefinitions of Pheromones

Most scientists would agree that for terms to have scientific utility, they must be operationally defined, as exemplified by consensual definitions for such agents as hormones, neurotransmitters, and neutrons. But what, in fact, are pheromones? How are they to be recognized or identified? It is instructive to see how authoritative sources, such as dictionaries, textbooks, and scientists working in the field of odor communication, have defined pheromones and the degree to which such definitions are congruent with one another and how far they have strayed from the original, seemingly canonized, definition of Karlson and Lüscher (1959). Obviously there are hundreds, if not thousands, of such sources, so a few must suffice to make the general point.

Dictionary Definitions

The *Random House College Dictionary* (1975) defines pheromones as follows: "*n. Biochem.* any of a class of hormonal substances secreted by an individual and stimulating a physiological or behavioral response from an individual of the same species [< Gk phér(ein) (to) bear + -o- + (HOR)MONE]." By this definition, pheromones are equivalent to hormones and produce physiological or behavioral responses in a recipient of the same species. *Webster's New Collegiate Dictionary* (1999), on the other hand, does not specifically require a pheromone to be a hormone and confines the example of an elicited response to a behavioral one: "*n.* a chemical substance that is produced by an animal and serves esp. as a stimulus to other individuals of the same species for one or more behavioral responses." By this definition, the animal must produce the chemical, not simply serve as an intermediary in its transport. *Dorland's Illustrated Medical Dictionary* (1981) similarly uses a less restrictive definition: "a substance secreted to the outside of the body by an individual and perceived (as by smell) by a second individual of the same species, releasing a *specific reaction or behavior* [italics mine] in the percipient." Again, an endocrine influence is not explicitly noted, although perhaps evoking the concept perception implies mental awareness on the part of the recipient. Species specificity, as well as a circumscribed reaction or behavior, is indicated. In contrast, *Stedman's Medical Dictionary* (1999) employs a more traditional hormone-based definition, although perception is again required and the general nature of the response is noted: "A type of ectohormone secreted by an individual and perceived by a second individual of the same species, thereby producing a change in the sexual or social behavior of that individual." *Webster's New World Medical* Dictionary (2008) is more explicit in defining pheromones as hormones and introduces the concept of volatility: "Pheromone: An agent secreted by an individual that produces a change in the sexual or social behavior of another individual of the same species; a volatile hormone that acts as a behavior-altering agent." Oxford University Press's *Dictionary of Science* (1999) indicates that pheromones are ectohormones that serve as signals to individuals "usually of the same species." Specific mention is made of the nature of the typical chemicals involved and their explicit presence in mammals: "pheromone (ectohormone): A chemical substance emitted by an organism into the environment as a specific signal to another organism, usually of the same species. Pheromones play an important role in the social behavior of certain animals, especially insects and mammals. They are used to attract mates, to mark trails, and to promote social cohesion and coordination in colonies. Pheromones are usually highly volatile organic acids or alcohols and can be effective at minute concentrations."

It is noteworthy that none of these definitions restrict their application to invertebrates and that pheromones are commonly viewed as externally secreted hormones. By implication the term *pheromone* is to be applied to a broad spectrum of animals. Only one of these definitions implies that pheromones are mediated by the olfactory system, albeit nonexclusively, although the need for pheromones to be perceived or to be volatile, as in some definitions, may convey this implicitly. Species specificity seems to be the most common generally held dictionary requirement of a pheromone.

Textbook Definitions

In contrast to dictionaries, textbook definitions seem to be more likely to overtly or covertly imply that a pheromone is a type of olfactory stimulus. As with dictionaries, some textbooks have a brief and simple definition of pheromones, whereas others provide more elaboration. At one extreme is the simple definition provided by Bear et al.'s basic neuroscience textbook (Bear et al., 2006), where a pheromone is defined as "an olfactory stimulus used for chemical communication between individuals" (p. 809). This definition is quite similar to that of Konrad Lorenz's student, Irenaeus Eibl-Eibesfeldt, who defines pheromones in his classic 1970 textbook, *Ethology*, as being species-specific without the requirement of olfaction being explicitly involved (Eibl-Eibesfeldt, 1970: "[Chemical] substances that are effective in intraspecific communication are called *pheromones*" [p. 69]). In both of these cases, pheromones are not explicitly equated to hormones. Groves and Schlesinger (1979), in their *Introduction to Biological Psychology,* also do not equate pheromones with hormones, although they, like Bear et al., assume pheromones require olfactory mediation. Furthermore, they state that pheromone receptors are highly specialized and that humans have pheromones: "Many animals and insects use odors to mark territories, to attract mates, to signal sexual receptivity, and for a variety of other communications. In many instances, the receptors for these chemical communicators are highly specific and respond to remarkably low concentrations of them. These chemical messengers are termed pheromones and are used by many organisms including human beings" (p. 293). In contrast, the hormone-like nature of pheromones is emphasized by Campbell (1996) in the fourth edition of his general biology textbook: "Pheromones are chemical signals that function much like hormones, with one important exception; instead of coordinating the parts of a single animal's body, pheromones are communication signals between animals of the same species" (p. 914).

Some authors equate pheromones with odorants coming from specific glands separate from urine or other excretions. For example, Dember and Jenkins (1970), in their introductory psychology textbook, note the following: "One special class of odorants has been of especial interest recently to biologists and animal psychologists. These are odorous substances secreted by an animal from special glands and left as an 'odor trail' wherever the animal goes. Such substances are called *pheromones;* they are known to be secreted by a wide variety of species, ranging from ants to gerbils. Analogous to pheromones in function, if not exactly like them physiologically, are odorous substances, such as urine, excreted by animals and left at various places in the territory over which they travel. One function of pheromones and similar odorants is to 'mark off' an animal's territory... Another function of such odorants is that of a trail-marker" (p. 282).

In a contemporary text focused on pheromones entitled *Pheromones and Animal Behaviour: Communication by Smell and Taste,* Wyatt (2008) indicates the following: "Pheromones are the molecules used for communication between animals... Strictly speaking, pheromones are a subclass of semiochemicals, used for communication within the species (intraspecific chemical signals)" (p. 1). Nevertheless, Wyatt takes a more heuristic approach in describing the influences of chemicals on behavior, stating,

"I have taken a broad and generous approach that includes many examples of behaviours mediated or influenced by chemical cues that would currently fall outside a rigid definition of pheromone" (p. 2). As will become apparent throughout this volume, such a liberal view seems necessary to accurately describe most chemically mediated mammalian behaviors and endocrine responses attributed to pheromones.

Definitions of Working Scientists

It is apparent that pheromones have not been uniformly defined by either dictionaries or textbooks. That being said, one would hope that scientists working in the field of chemical communication would employ consistent definitions of the term *pheromone*. This seems to be the case for many of those who work with non-mammalian organisms, where the equivalence of pheromones to externally secreted hormones is maintained. A case in point is the 1987 review of pheromones in algae by Maier and Müller: "The term 'pheromone' was introduced in 1959 by Karlson and Lüscher to characterize substances that coordinate activities between individuals within one species (Karlson and Lüscher, 1959). Such substances can be subsumed under the term 'hormone' in its broadest definition, but it is convenient to refer to diffusible substances traveling through the external media of air or water by using the more specific term 'pheromone'" (p. 146).

However, not all scientists, even those working with invertebrates and non-mammalian vertebrates, have confined the term to its implicit hormone-like meaning, although elements of the original definition are usually maintained. Shorey (1976), in a book entitled *Animal Communication by Pheromones*, defines pheromones as follows: "Pheromones are either odors or taste substances that are released by organisms into the environment, where they serve as messages to others of the same species" (p. vii). Signoret (1976) notes that "the term 'pheromone' has been widely used for any substance produced by an individual which, on contact with a member of the same species, evokes behavioral and/or physiological responses. Theoretically, such substances may act not only by olfaction but also by ingestion. In mammals, however, they are believed to act mainly via the sense of smell" (p. 243). Aron (1979) follows the original definition of Karlson and Lüscher (1959), but emphasizes olfaction, stating, "A sensory stimulus whose action is prevented by either olfactory bulb removal or peripheral anosmia should be considered a pheromone in nature until we obtain further information on the compounds involved" (p. 230), a position taken earlier by Rottman and Snowdon (1972). Izard and Vandenbergh (1982) employ a simple definition of pheromone that seems to maintain the concept of species specificity and the elicitation of a specific response but is rather nondescript: "Pheromones are chemical messages secreted by one animal that cause a specific reaction in another individual of the same species" (p. 189). Meredith (1998), moving afield from traditional definitions, defines pheromones as "chemicals used for mutually beneficial chemosensory communication between members of a species" (p. 349). This redefinition, which is reminiscent of a proposal by Rutowski (1981), has some attractive attributes, although a number of phenomena traditionally assumed to be mediated by pheromones may not provide *mutual* benefit, which itself is difficult to establish operationally (e.g., the elicitation of agonistic responses that may decrease the mating success of one participant over the other).[3]

In an introduction to a bibliography on mammalian pheromones, Whitten and Champlin (1972) provide a state-of-the-art description of what pheromones were believed to be in the early 1970s and add the observation that chemicals that are truly pheromones are most likely found in only one sex. They also suggest that some chemicals are "co-pheromones" and replace the term *releaser pheromone* with *behavioral pheromone*.

> [Pheromones are] substances, produced by one member of a species, which influence other members of the same species. They may be *behavioural pheromones* and evoke rapid behavioural reactions through nervous pathways or they may be *primer pheromones* and induce relatively slow endocrine responses. There will, of course, be some overlap in this classification because behavioural responses may eventually follow endocrine changes and some responses such as alarm and stress will involve neuroendocrine pathways and may be intermediate in time sequence...
>
> Identification of an odorous substance from a glandular or other secretion does not prove that the substance is a pheromone. If it is limited to one sex as, for example, muscone in the pod of the male musk deer, then the case is considerably stronger. *If, however, as is reported for civetone*

it is produced by both sexes and by related species then one must look for another function [italics mine]. Civetone has been used for centuries as a fixative in perfume manufacture. For this purpose it intensifies and prolongs more subtle and ephemeral components of a perfume. Civetone may perform this same function for the civet and perhaps it could be considered a co-pheromone. (p. 150)

A number of relatively recent definitions of pheromone by scientists are worthy of note. Jacob et al. (2004) state that "pheromones are defined as those natural compounds produced by one member of a social group that can regulate the neuroendocrine mechanisms underlying fertility, development or behavior of another group member" (p. 422). This definition raises the question as to what comprises a "natural compound" and replaces *species* with the general term *social group*. Stern and McClintock (1998), in a study claiming the demonstration of the first human pheromone (which, in fact, was not chemically identified), reiterate the belief that pheromones are airborne and, thus, volatile: "Pheromones are airborne chemical signals that are released by an individual into the environment and which affect the physiology or behaviour of other members of the same species" (p. 177). This definition, like other definitions requiring volatility, would exclude many substances previously deemed pheromones in aquatic environments from the pheromone class, as well as nonvolatile agents detected upon contact by ingestion, the vomeronasal organ, or other means. These workers further indicate that "here we investigate whether humans produce compounds that regulate a specific neuroendocrine mechanism in other people *without being consciously detected* [italics mine] as odours (thereby fulfilling the classic pheromone definition)" (p. 177) It is not clear what "classic" definition of pheromone is the referent, although as will be noted later in this chapter, workers such as Singer, Axel, and others have expressed the belief that pheromones may be unconsciously mediated by the vomeronasal organ (Singer, 1991; Dulac and Axel, 1998; Belluscio et al., 1999). Carrying the notion of volatility even further, Savic et al. (2001) incorrectly note that "the pheromones are, according to the original [i.e., Karlson & Lüscher] definition, *volatile* [italics mine] compounds secreted into the environment (in sweat, urine) by one individual of a species." These authors go on to state that an important criterion for an agent to qualify as a human pheromone is its ability to "activate the human hypothalamus in a sex-specific mode" (p. 661). Recently, Martinez-Ricos et al. (2007) stated, "We assume that these substances [in mouse urine] are involved in intersexual communication because they elicit reliable responses (attraction and reinforcement) that have a strong influence on the behavior of adult females. Therefore they fulfill the defining features of sexual pheromones" (p. 145).

Clearly, as with the case of textbooks, quite different perspectives are held by scientists as to what constitutes a pheromone. It is apparent, however, from the writings of those working in the chemical communication field, as well as in dictionary and textbook definitions, that the presumptive effects of pheromones are largely independent of experience. Such a perspective would seem necessary if pheromones are to be distinguished from other chemicals to which conditioning can easily occur. This follows logically from the insect origin of the pheromone concept and its generalization to mammals. Thus, Buck (2000) states that "pheromones elicit programmed neuroendocrine changes and innate behaviors" (p. 611). Reynierse (1974) indicates that it is necessary for "the activities of both the sender and receiver [to be] unconditioned patterns of behavior" (p. 4) for a substance to be a pheromone. Unfortunately, the distinction between innate and learned responses is not easy to maintain. As acknowledged by Whitten and Champlin (1973) in relation to putative pheromones involved in sexual and reproductive processes, it is "difficult to distinguish experimentally between pheromones and other, perhaps trivial, odors that have previously been associated with sexual rewards" (p. 4).

Early Concerns with the Concept of Releasing Pheromones in Mammals

The use of the *pheromone* term for stimuli that alter endocrine state, that is, so-called primer pheromones, has rarely been questioned although, as noted in Chapter 6, it is clear that this neglect is not warranted. In contrast, as early as the late 1960s a number of influential biologists voiced concern about the utility of describing chemicals involved in the social behavior of mammals as "releasing" pheromones, reflecting their awareness that mammalian behavior is not reflexive in the same way as the behavior of many invertebrates. Bronson (1976a) articulated the problem as he saw it more than 30 years ago:

> It is perhaps unfortunate that interest in mammalian chemical communication blossomed at a time when the study of insect pheromones was already a sophisticated field of research. Thus Whitten (1966) introduced the primer-releaser dichotomy to mammalian workers and Bronson (1968) amended it only slightly by arguing that the term signaling was a more appropriate modifier than releaser for a nonprimer pheromone, given the variable, experience-oriented behavior of mammals. The unfortunate side of such generalizations is the tendency to think of mammalian communication in terms of simple stimulus-response systems. For example, it is now relatively common usage to refer to "aggression-promoting (or eliciting)" and "aggression-inhibiting" pheromones in mice (e.g., Lee and Griffo, 1974; Mugford and Nowell, 1972). The obvious implication of this terminology is the existence of two simple urinary compounds which unequivocally either release or inhibit a stereotyped aggressive response. Mammalian social behavior simply does not work that way except at the purely reflexive level.[4] (p. 123)

In the same book in which Bronson expressed his concerns about the concept of releasing pheromones, Beauchamp et al. (1976) questioned "the current usefulness of the term 'pheromone' in describing the influences of biological secretions and excretions upon mammalian reproductive behaviors" and suggested "that the uncritical use of this term has led to a number of misconceptions in the interpretation and conduct of mammalian behavioral research" (p. 144). These workers compiled a list of what the implicit or explicit criteria for a chemical to be termed a pheromone seemed to be up to that time—a list that at least provided an operational basis for determining whether a chemical was, in fact, a pheromone. The criteria were as follows:

- species specificity
- a well-defined behavioral or endocrinological function
- a large degree of genetic programming
- the involvement of only one, or at most, a few compounds

- uniqueness of the isolated compounds or small set of compounds in producing the behavioral or endocrinological response

All existing claims of the isolation of mammalian pheromones were evaluated. Beauchamp et al. pointed out that none had been tested for species specificity or had met even half of these criteria. Indeed, only one criterion—that of assumed chemical simplicity for the isolated product—was met by all of the isolated substances. These authors concluded, "It would appear to us that the labeling of a compound as a pheromone, when it has not been demonstrated to meet a well-defined set of operational criteria, is problematic if the pheromone term is to have any meaning beyond that of being synonymous with a 'chemical'" (p. 148).

This early critique of the pheromone concept aroused considerable debate and consternation, with one well-established scientist withdrawing his chapter from the to-be-published volume after learning of the anti-pheromone chapter and another accusing the authors of jeopardizing his funding. Some entomologists came forward with sharp criticism (Katz and Shorey, 1979), which evoked rebuttal (Beauchamp et al., 1979). Booth (1980), whose research mainly focused on mammals, argued the following:

> There have been reservations concerning the suitability of the original pheromone concept when applied to chemical communication in mammals. In mammals it appears that many behavioural and physiological responses to pheromones are no more specific than responses to hormones; for example, both androgens and oestrogens can induce male mounting behaviour and produce anabolic effects. Thus in relation to mammalian physiology the term pheromone is just as valid and convenient to use in the study of chemical communication (exocrinology) as the term hormone in endocrinology. (p. 307)

Booth's argument is weak on a number of grounds, not the least of which is the fact that only those androgens that can be metabolically converted to estrogens induce male mounting behaviors. While it is true that behavioral responses to hormones depend upon such factors as prior hormone exposure, season, and the age of the organism, hormonal effects are generally robust and reliable (Pfaff et al., 2004). In contrast, responses of many mammals to putative pheromones are highly variable, in some cases failing to induce the intended response in most of the subjects under study (see Chapters 5 and 6). Hormones are rarely species-specific in mammals and, unlike putative pheromones, most have been identified and have a limited number of well-defined cognate receptors. Booth's argument fails to address the issue of learning, although he did state that "it seems that, whereas the response to pheromones in insects is primarily innate, in mammals learning by association with other stimuli can be involved" (p. 303).

The issue of learning, addressed in detail in Chapter 3, is critical to acceptance of whether putative mammalian pheromones differ from odorants in general. This problem was known and acknowledged by a number of the pioneer mammalogists in this field, some of whom expressed reservations about the validity and viability of the releaser pheromone concept in mammals. For example, Bronson (1976a) wrote:

> Experience is a profound modifier to mammalian social behavior. There have actually been relatively few attempts to examine the role of experience in odor-induced responses in mammals. Where investigated, however, the results usually have indicated a potent role for experience. Thus species identification apparently can be easily manipulated by odors early in the life of mammals (e.g., Carter and Marr, 1970; Mainardi et al., 1965; Marr and Lilliston, 1969) and adult sexual experience is a strong determinant of response to sex odors (e.g., Caroom and Bronson, 1971; Carr et al., 1965; Carr et al., 1966). One wonders at this point whether the pheromone concept, so useful in insect behavior and physiology, should be bastardized to the point where it is used to cover situations in mammalian behavior where usually complex odors evoke highly variable responses which are easily modified by experience. (p. 123)

In light of this issue, a number of investigators sought to develop additional classifications of pheromones that would take into account the importance of learning. For example, Müller-Schwarze (1977) suggested that "informer" pheromones should be added to the pheromone lexicon since some chemical signals in mammals are "stored in the memory and can be recalled later in a variety of contexts" (p. 421). Hilda Bruce, one of the pioneers of early mouse pheromone research, suggested the term *imprinting pheromones* be used for chemicals in the nursing setting that influence adult behavior (Bruce, 1970). Her suggestion stemmed from the following observations:

> In mice (Mainardi et al., 1965) and in rats (Marr and Gardner, 1965), social behaviour in the adult may be modified by olfactory experience during the period of suckling. Female mice reared in the absence of the father show a loss of discrimination in sexual selection when adult, and the same deficiency develops if the olfactory atmosphere of the nest is artificially altered by spraying the parents every day with perfume (Parma violet). Young rats reared from birth to about four weeks of age in the olfactorily artificial atmosphere, scented either with Yardley's Red Roses cologne or with oil of wintergreen, also showed modifications when adult. (pp. 12–13)

Recent research, described in more detail in Chapters 4 and 5, suggests that the situation is more complex than appears on the surface (Moncho-Bogani et al., 2002, 2005). According to this work, female mice that have been isolated from male mice and their odors from the time of adolescence show no preference for male mouse urine odor. However, after an initial vomeronasal organ–dependent encounter with male soiled bedding, an encounter which is said to be intrinsically rewarding, a rapid conditioning occurs to the odor of the urinary volatiles which is mediated by the main olfactory system. Interestingly, if the urine stimulus to which the female is initially exposed comes from a group of males, the preference generalizes to a number of males. However, if the exposure is to urine from a single male, the preference is specific to that male's urine and no preference is shown for urine odor from other males (Ramm et al., 2008).

The Chemical Complexity Issue

In accord with the original insect model and the desire to isolate "the pheromone" responsible for a given behavior or endocrine response, most investigators have assumed that mammalian pheromones consist of one or at most a few chemicals that induce defined responses. This premise stems in part from the successful 20-year-long effort to identify the silkworm moth sex attractant bombykol (Butenandt et al., 1959). It should be emphasized, however, that mammalian secretions and excretions are typically made up of hundreds of volatile and nonvolatile agents. Moreover, from a theoretical perspective, single molecules, compared to mixtures, have significant limitations as to the amount and distinctiveness of the information they can convey. While large molecules, particularly ones that can take on several different conformations, could convey considerable information, their synthesis is energetically expensive and, if volatility is involved, size limitations are present. Hence, mixtures of relatively small compounds have advantages in terms of signal diversity, information transfer, and flexibility of synthesis over single chemicals which, if much information is to be transferred, must become structurally complicated.

In light of the chemical complexity of mammalian secretions and excretions, an important question arises: When does the number of involved chemicals move from the pheromonal to the non-pheromonal? Johnston (2000) proposes that the term *chemical signal* be used as a generic term for chemical compounds or mixtures released into the environment that serve intraspecific behavioral or physiological functions. In accord with the core assumption of the pheromone concept, he reserves the term *pheromone* for chemical signals that employ a single compound and the term *pheromone blend,* which has been employed in some insect studies, for mixtures of a small number of compounds that are maximally effective when they occur in precise ratios. He further suggests that the term *mosaic signal* or *odor mosaic* be used to refer to mixtures of large numbers of compounds in which many components are important for producing the full effect. Johnston notes that "this scheme has the advantage of preserving the traditional use of the terms 'pheromone' and 'pheromone blend,' while adding a third important category to encompass signals that have previously been given little attention or regulated to a second-class status" (p. 104). He makes no distinction between learned and unlearned responses to the chemicals in question.

Unfortunately, defining mammalian pheromones on the basis of the number of involved chemicals becomes difficult since theoretically the more chemicals that are added to a mixture in appropriate proportions, the closer the mixture will come to mimic the effects of the original secretion. However, when volatility is involved, some chemicals alter the vapor pressures of other chemicals in complex ways and the degree to which a given chemical volatizes or enters the mixture can depend upon such factors as proteins to which it is bound and its propensity to interact with other chemicals in the mixture. Some chemicals can serve as antagonists to one another at the receptor level, and it is well known that vapor pressures of mixtures fail to follow predictions from Raoult's Law (Haring, 1974). Hence, the order in which chemicals are added to a mixture can alter a mixture's characteristics and, in some cases, many chemicals must be added to the mixture to closely mimic the parent secretion or excretion.

In light of such issues and the general failure to identify unique single molecules or small sets of molecules that fully replicate the effects of the parent stimuli (see Chapters 5 and 6), it is not clear what advantage exists in maintaining a concept in mammals that implicitly or explicitly defines the nature of the interaction as hormone-like or fundamentally different from other chemically mediated phenomena.

Pheromones as Non-odors

In the late 1980s, A.G. Singer, the leader of a group that had performed the most analytical work on the search for mammalian pheromones, pointed out the lack of progress made even by that time in identifying such entities (Singer et al., 1987):

> When our group began work in this area there was very little chemical evidence for mammalian pheromones, that is for single compounds or simple mixtures that acted like hormones between the individuals of a population to coordinate functions, such as reproduction, vital to the survival of the population. A number of pheromones had been characterized in various insects, but much of the chemical work in mammals, particularly the analysis of volatile compounds from scent glands or urine, led only to the conclusion that chemical communication in mammals is complex. (p. 287)

In 1991 Singer set the stage for more contemporary theorists in arguing that pheromones may, in fact, not be odorants. He also countered the concept that volatility is a critical element of pheromones:

> It is possible that compounds with potent pheromonal effects have no particular odor. As a consequence of this reasoning, pheromones in mam-

mals should not necessarily be considered as a subclass of social odors. They may be more usefully considered to constitute a separate class of biologically active compounds with a function of regulation, rather than transfer of information...

When we look for mammalian behavioral and physiological responses to pheromones in which the responses appear to be specific, we find that a remarkable number of them are mediated by accessory olfactory organs such as the vomeronasal organ rather than by primary olfaction, and that many of the pheromones are transferred by contact (Singer et al., 1987). Since the stimuli are not necessarily olfactory, we have to consider the possibility that they are nonvolatile, that the pheromones are not necessarily airborne, but may be transferred by direct contact of the responding animal with the stimulus source. With the removal of the restriction to volatile compounds, even macromolecules such as proteins, could function as pheromones, which is what we have found in our identification of an aphrodisiac pheromone in hamsters. (p. 628)

Although volatility was not a key element of most prior pheromone definitions, volatility is implicit in definitions that suggest pheromones are odorants mediated via the main olfactory system. Thus, Singer, particularly within the context of analytical attempts to identify pheromones, expanded the nature of stimuli that were fair game for pheromone hunters. However, as shown later in this volume, such expansion has not led to a plethora of identified pheromones. Indeed, it is questionable whether the putative hamster aphrodisiac pheromone mentioned by Singer accurately mimics the parent secretion, as described in Chapter 6.

The Vomeronasal Organ as the Pheromone Receptor System

Subsequent to Singer's arguments that nonvolatile odorless agents could serve as pheromones, a number of molecular biologists proposed the vomeronasal organ (VNO) as "the pheromone receptor," relegating the main olfactory system to the perception of non-pheromones (Belluscio et al., 1999; Dulac and Axel, 1998; Tirindelli et al., 1998; Buck, 2000). This tube-like structure, also known as Jacobson's organ, is the peripheral receptive element of the accessory olfactory system and is located bilater-

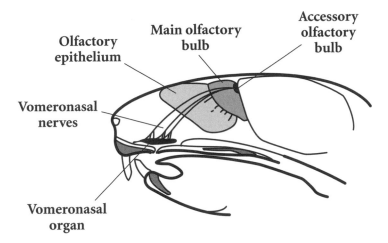

Olfactory epithelium

Main olfactory bulb

Accessory olfactory bulb

Vomeronasal nerves

Vomeronasal organ

FIGURE 2.3. Sagittal view of the rodent vomeronasal organ system showing the location of the tubular vomeronasal organ, vomeronasal nerves, and accessory olfactory bulb in relation to the main olfactory system. Modified from Brennan and Keverne (2004).

ally at the base of the nasal chambers in most mammals, reptiles, and amphibia (Wysocki, 1979) (Figure 2.3). The central projections of this organ are separate from those of the main olfactory bulb, with afferents first going to the accessory olfactory bulb and then to the hypothalamus via the medial amygdala. In mammals, pump-like engorgement and disengorgement of blood vessels and sinuses lining the VNO's lateral wall bring materials into its lumen from small ducts connected to the anterior part of the nasal cavity, or in some species, the oral cavity (Meredith et al., 1980). Its receptors are located on microvillae, rather than cilia (Døving and Trotier, 1998).

The two primary gene families that express the G-protein-linked VNO receptors are only distally related to the multigene family that expresses most of the receptors in the main olfactory system, implying that the VNO responds to somewhat different classes of stimuli. However, the VNO also expresses more than 40 receptor genes of the type found in the main olfactory epithelium, with the expressing cells projecting to the accessory, not main, olfactory bulb (Levai et al., 2006). The cells within the VNO and the accessory olfactory bulb typically respond more slowly and sustain longer bouts of firing than those within the main olfactory epithelium, suggesting to some that the VNO is not specialized for behaviors that require rapid responding (Luo et al., 2003). Nonetheless, this organ plays a significant role in individual recognition and agonistic behavior. In some vertebrates, including many reptiles, the VNO is involved in food finding and predator and prey identification, in addition to reproduction. Although humans have rudimentary VNO sacs on each side of the anterior septum that connect to the nasal cavity via bilateral pits located 1 to 2 cm from the posterior margin of the naris (Moran et al., 1991), this structure lacks the full complement of neural elements necessary for function, as described in detail in Chapter 7.

The concept of the VNO as the pheromone receptor is well illustrated by this passage from Dulac and Axel (1995):

> Mammals possess an olfactory system of enormous discriminatory power. Humans, for example, are capable of recognizing thousands of discrete odors. The perception of odors in humans is often viewed as an aesthetic sense, a sense capable of evoking emotion and memory, leading to measured thoughts and behaviors. Smell, however, is also the primal sense. In most species, odors can elicit innate and stereotyped behaviors that are likely to result from the nonconscious perception of odors. These different pathways of olfactory sensory processing are thought to be mediated by two anatomically and functionally distinct olfactory sensory organs, the main olfactory epithelium (MOE) and the vomeronasal organ (VNO). (p. 195)

Unfortunately, this concept has many problems. For example, removal of the VNO does not meaningfully affect a number of behaviors traditionally viewed as being influenced by pheromones, such as the mating behavior of the guinea pig (Beauchamp et al., 1982) and the lordosis response of sows to boar odors (Dorries et al., 1997). Importantly, the VNO responds both *in vivo* and *in vitro* to chemicals not generally viewed as pheromones (Hatanaka et al., 1988; Hatanaka, 1992; Meredith, 1980; Luo et al., 2003; Sam et al., 2001),[5] limiting the notion of its specificity to pheromones. Deficiencies in the TRP2 cation channel, which is critical for VNO function, results in insensitivity to a range of chemical stimuli, not just those that have been deemed pheromones.[6]

Such observations, along with the fact that "both the main and accessory olfactory systems have access to neural systems controlling endocrine and behavioral responses" (Zufall and Leinders-Zufall, 2007, p. 486), have led some to revise their concept of the VNO as the pheromone detector. For example, Axel's group now states that "mammals have evolved innate behavioral arrays mediated by pheromones that activate both the main and the olfactory system" (Leypold et al., 2002, p. 6376). In a 2005 review entitled "What Is a Pheromone? Mammalian Pheromones Reconsidered," Stowers and Marton indicate that "recent studies in conjunction with prior evidence suggest that the working definition of pheromones as nonvolatile molecules that regulate innate social behavior by activating vomeronasal organ (VNO) sensory neurons may be too restrictive. Indeed, it appears that pheromones may be nonvolatile or ephemeral, activate VNO or main olfactory epithelium (MOE) neurons, and may have their effects altered by context as opposed to being strictly innate" (p. 699).

In aggregate, the aforementioned observations strongly suggest that pheromones cannot be distinguished from non-pheromones on the basis of VNO activation, negating the alluring clarity of this distinction. Nonetheless, the VNO does differ from the main olfactory system in several interesting ways, most notably in its receptor repertoire and its central neural connections.

Pheromones and Politics

Millions of dollars have been spent by the perfume industry and the federal government over the past half century in the largely unsuccessful attempt to identify mammalian pheromones, making their existence a hot political and economic issue. The popular belief in pheromones is widespread, leading to the development of lotions, perfumes, and personal care products containing putative pheromones, a number of which purportedly increase sexual attractiveness and instill greater confidence in many questing to attract mates (see Chapter 7). This popularity is reflected by the fact that over five million hits, most of which are related to human pheromones, occur when one Googles the term *pheromone.*

As pointed out earlier in this chapter, some scientists feared in the mid-1970s that their grant funding would be jeopardized by the questioning of the validity of the pheromone concept in mammals. Although such fears were not realized, the politics of this issue remains today, particularly in light of the emphasis on granting agencies to fund translational research that ultimately benefits humans. Lundström (2005) recently stated the following in an argument for maintaining the term *pheromone* in studies of human chemical communication:

> [The fact that humans use chemicals in communication] renders the heated debate on the existence of human pheromones, although important in itself, more of a semantic character. However, what should the demonstrated effects be termed if not pheromonal?
>
> Several suggestions have been put forth such as semiochemicals, ectohormones, vomeropherins, to mention only a few. I argue that the use of the term "pheromone" is still appropriate. The word and its definition (although not agreed upon in detail) are well known among scientists and the laymen alike. *Moreover, the word has an attraction in itself that promotes the field in a way that no other word can do* [italics mine]. (p. 53)

Eight years ago, when the vomeronasal organ was being touted as the pheromone receptor, Beauchamp (2000) dissociated himself from his 1976 critique of the pheromone concept, indicating that perhaps *the most important reason* for *not* abandoning the use of the term *pheromone* in mammals is that

> the term provides a useful and powerful marketing tool for good—as well as bad—science. It is much more sexy to study human (or even non-human) pheromones rather than to study chemosignals or chemical communication in humans. This is true whether the marketing target is the mass audience of newspaper, television or the internet, or a more restricted group such as grant reviewers. (p. 2)

He goes on to state,

> definitions that are too specific can become straight-jackets. So now, when I am asked whether there are human pheromones I say: of course. But that is only the start of the conversation and of the research. What we really want to know is what is the chemistry involved, how is physiology effected, and how this is translated into a behavioral response. These are the same questions we had 25 years ago. Now, with more powerful tools both

in chemistry and biology, we are in the process of getting much better answers. (p. 2)

Despite such optimism, there is little evidence for human pheromones and answers to such questions are unlikely to be forthcoming, as will be shown in Chapter 7.

THREE

Mammals Are Not Insects

By themselves, odors suggest nothing. I must learn by association to judge them of distance, of place, and of the actions or the surroundings which are the usual occasions for them, just as I am told people judge from color, light, and sound.

Helen Keller, *The World I Live In,* 1908

The problems with the pheromone concept described in Chapter 2 are not generally appreciated. Mammals have much more complex nervous systems than insects. A number of their physiological responses to external stimuli—including overt behaviors and internal hormonal changes—often are tempered, guided, and in some cases determined by such intervening factors as context, stress, learning, and memory of past experiences. This is not to denigrate the remarkable ability of insects to learn (Dukas, 2008) and the fact that many insect behaviors are complex and dependent upon multimodal signals, as will be discussed in Chapter 8, only to state that the mammalian repertoire of endocrine and behavioral responses is seemingly more elaborated and less stereotyped.

In this and the following chapter, I describe how learning is critically important for establishing the meaning of the vast majority of biologically based chemicals in mammals. I provide examples that show that the olfactory system, in many cases, is not fundamentally different from other sensory systems in terms of its influences on behavior and endocrine function. For example, odorants present in early life significantly impact social behavior later in life in a manner analogous to the way that visual and auditory stimuli, via "imprinting," influence later social behaviors of many birds and some other vertebrates. Moreover, I point out that alterations of reproductive processes of rodents induced by repeated or regular presentation of male urine can also be induced by repeated presentation of loud sounds and stress-inducing lights and that the olfactory system cannot be divorced from ideation and emotionally laden experiences that, in turn, can alter reproductive states. In fact the anatomical associations of the olfactory system provide a rich substrate for the cognitive mediation of complex odor-related phenomena, as noted by Slotnick (2002):

> Recent discoveries have revealed that the olfactory system is less simple and less primitive than is generally assumed: olfactory impulses have fairly direct inputs to brain regions implicated in complex functions, including limbic structures and the prefrontal cortex...
>
> The first link between olfaction and cognition was the finding that cells in the olfactory cortex project to the segment of the thalamic mediodorsal nucleus that connects to the orbital frontal cortex. Subsequent reports confirmed the existence of an "olfactory thalamocortical circuit," and delineated olfactory connections to the amygdala, entorhinal cortex and hypothalamus. (p. 216)

In the 1960s, the comparative intelligence of a number of animals was assessed using "learning sets" or "learning to learn" paradigms based upon visual responses, one of several approaches for assessing animal cognition. In such studies, the test subject is given a series of discrimination problems to solve, the first of which often requires many trials to learn. Over a series of sessions, however, the ability to solve new problems dramatically improves, suggesting that the animal has learned "rules" or "concepts" underlying the task, such as "oddity."

A number of studies of that era had suggested that the relative performance of various species on such tasks mirrored the "phylogenetic scale," or *scala naturae,* employed, often inappropriately, in studies of comparative anatomy (i.e., primates → other mammals → birds → reptiles → amphibians → fish → insects [Bitterman, 1965]; for critique, see Hodos and Campbell, 1969). However, when the sensory specializations of different forms were taken into account, it became apparent that rodent species such as the rat could perform essentially as well as primates on such tasks, so long as olfactory, rather than visual, stimuli were employed. Thus, Jennings and Keffer (1969) and Nigrosh et al. (1975) found excellent interproblem transfer over a series of two-odor discrimination problems in the rat, such that errorless performance on subsequent reversals was commonly attained (Slotnick, 2000). These and other studies demonstrated that rats can learn and discriminate among large numbers of odors, and can remember whether they were reinforced or not reinforced for each of these odors in a test series (Slotnick et al., 1991). Slotnick (2001) stated:

> Functional studies have overcome many of the technical difficulties of controlling vapor stimuli and demonstrate that, with odor cues, rats display highly efficient learning rivaling that of primates. In short, the evidence indicates that rats can "think with their noses" and have the neural machinery to do so. This evidence, combined with advances in the molecular biology of olfaction (Mombaerts et al., 1996), has resulted in a renaissance in research on olfaction and to the surprising and occasionally controversial suggestion that the rodent olfactory system could serve as a model for neurobiological studies of cognition (Reid and Morris, 1993; Slotnick, 1994). (p. 216)

An inherent problem with the pheromone concept when applied to mammals is that, by implication and definition, it obfuscates or excludes from consideration the possibility that cognitive processes are involved in the mediation of chemically induced behavioral and endocrine responses. By focusing primarily on the stimulus, it divorces from the communication and endocrine processes important influences from greater regions of the brain. Most would agree that nonhuman mammals have ideation or thoughts, such as occur during episodes of REM (rapid eye movement) sleep dreaming when they exhibit sniffing and other seemingly odor-related behaviors (Rasmussen et al., 1993). Enhanced secretion of luteinizing hormone and testosterone occurs in rats, cattle, sheep, and other animals in anticipation of sexual activity (Graham and Desjardins, 1980). In men, testosterone titer increases in winners and decreases in losers of tennis, wrestling, debates, chess matches, and various games, a number of which are largely intellectual enterprises (Elias, 1981; Gonzalez-Bono et al., 1999; Suay et al., 1999; Mazur et al., 1992; Rejeski et al., 1989; McCaul et al., 1992). Erotic dreams, expectation of sexual encounters, and vicarious identification of male sports fans with a winning team can increase testosterone levels (Anonymous, 1970; Carani et al., 1990; Hellhammer et al., 1985). On the other hand, identification with a losing team can decrease testosterone levels (Bernhardt et al., 1998). Chronic stress and mental concerns or conflicts are known to alter endocrine function of both male and female mammals (Hatch et al., 1999; Fenster et al., 1999; Beaumont, 1982), and it is well established that stress, induced by a variety of procedures, activates their dopaminergic, noradrenergic, GABAergic, and endorphinergic systems (D'Amato and Cabib, 1987, 1990; Nakagawa et al., 1981; Yoneda et al., 1983). Hence, the influences of odors on hormones need not be the direct result of the stimulus, but the stimulus influences on ideation, memory, and emotions, in some cases via the autonomic nervous system. The degree of the effect commonly depends upon prior experience or lack of experience with the involved stimulus.

Chemosensory Learning

If mammalian pheromones are specific and innate hormone-like agents that differ from the plethora of other environmental chemicals involved in the mediation of social behaviors or endocrine responses, one would expect them to be little influenced by learning. One would also expect them to be more or less genetically fixed, a point made by numerous theorists, including Bronson in regards to releasing pheromones (see Chapter 2). However, as described in this chapter, this is rarely the case, and learning seems to play the salient role in determining the meaning of the vast majority of odorants, many of which have been termed pheromones, to mammals. This dilemma leads to problems for the pheromone concept, as articulated by Moncho-Bogani et al. (2002):

> As put forward by Beauchamp et al. (1976) and further discussed by Nyby et al. (1978), since learning might confer "pheromonal" properties to virtually every odorant, the classical concept of pheromone cannot be easily applied to mammalian intersexual communication. Therefore, it seems appropriate to restrict the concept of pheromone to those substances that have innate biological value for intraspecific communication. (p. 174)

Some have argued that an unlearned initial attraction to a biological-based chemical substance is based on a pheromone, whereas increased subsequent responsiveness to the substance reflects conditioning of additional elements of the chemical stimulus to the pheromone. This approach requires that a scent source be divided into pheromones and odorants, which is a daunting task, and begs the question as to whether the whole pheromone concept has operational value. Even if some chemical stimuli are inherently more preferable than others, as is quite likely, one does not have to infer that an innate pheromone is the basis for this preference or is essential for the initiation of exploration or sniffing of a scent. Is it necessary to infer that an innate touchomone is the basis for preferences of some types of tactile stimulation or for the initiation of scratching or licking of the integument?

A strong argument can be made that, just as mammalian photoreceptors have not evolved *specifically* for visually detecting mothers, fathers, or Ferrari automobiles (even though such detection is possible through learning), so too mammalian olfactory receptors have not evolved *specifically* to detect the odor of mothers, fathers, or the exhaust fumes of Ferrari automobiles. This lack of specificity extends to the smells of individual conspecifics, even though their long-term identification, like the visual detection of Ferraris, can occur as long as learning at some point intervenes, possibly in conjunction with vomeronasal activity. While the olfactory system, like the visual system, can provide information about the physical nature of the environment, the specificity of such information is largely dependent upon experience. It is noteworthy that olfactory detection thresholds of rats for perflorocarbons—agents never encountered during their phylogeny—are at the same level of magnitude as thresholds for many organic chemicals presumably encountered during ancestral evolution, reiterating the notion that evolution has not focused on the detection of specific object-related chemicals but on the provision of a flexible sensory system sensitive to even *de novo* chemicals (Marshall et al., 1981).

Evidence for the important role of olfactory learning in prenatal, neonatal, and adult mammals is described below and is elaborated on more specifically in later chapters. Learning dictates nearly all so-called releaser pheromone effects, as well as a number of primer pheromone effects including the classic Bruce effect (blockage of implantation by the presentation of a strange male's odor). Mammals learn such important information as the odor of their parents, species, offspring, social group (e.g., deme), fecund sexual partners, dominant or submissive conspecifics, and familiar conspecifics, the latter of which distinguishes them from strangers.

Olfactory Learning in the Uterus

The olfactory system of many mammals, including humans, is functional *in utero*. Intrauterine learning can occur and manifest itself in postpartum life. Evidence for prenatal function includes observations that premature human infants exhibit discriminative responses among low concentrations of odorants presented to them (Sarnat, 1978; Pihet et al., 1997) and rat fetuses, transferred from the abdominal cavity of their mothers into saline without interruption of the maternal blood supply, exhibit increased activity, altered heart rate, and facial wiping responses to odorants (Smotherman and Robinson, 1987, 1990).

Prenatal experience with odors can influence behavior later in life (Schaal et al., 1995; Hepper, 1988). Human fetuses learn odors related to their pregnant mothers' diets (Schaal et al., 2000), which can be reinforced in the nursing situation, where flavors ingested by the mother can be transmitted via the mother's milk (Galef and Sherry, 1973; Mennella and Beauchamp, 1991a, 1991b, 1996; Galef and Henderson, 1972). Rat pups exposed to citral *in utero* attach, postpartum, to washed citral-scented nipples and not to normal unwashed nipples (Pedersen and Blass, 1982). Offspring of pregnant rats receiving an infusion of an odorant into the amniotic fluid and made sick by lithium chloride injected into the mother avoid postnatally the odor to which they had been exposed (Smotherman, 1982; Stickrod et al., 1982). If no toxic agent is administered to the mother, then a postnatal preference for the prenatally exposed odorant may appear in later life, particularly if that same odorant is present in the early perinatal period (Pedersen et al., 1983).

Intrauterine exposure to odorants can have long-lasting influences on olfactory neurophysiology. In one study, rabbit kits whose mothers had been fed juniper berries during pregnancy (such berries are part of the rabbits' natural diet) prefer red juniper long after weaning, even if raised by a foster mother fed standard laboratory food in the absence of subsequent experience with juniper (Bilko et al., 1994; Hudson and Distel, 1999). As adults these rabbits had a larger magnitude of the summated electrical potential at the surface of the olfactory epithelium in response to juniper than did adult rabbits whose mothers were not fed the berries, suggesting that long-lasting neural changes were induced in the olfactory receptors (Hudson and Distel, 1999). This observation is in accord with other studies demonstrating exposure-related alterations in both peripheral and central olfactory physiology (Wang et al., 1993; Youngentob and Kent, 1995; Mouly et al., 2001).

Olfactory Learning in the Nursing Setting

Contrary to the pheromone concept, and as mentioned earlier in this chapter, many odors are learned during early periods of the developing mammal in a manner analogous to the so-called visual and auditory imprinting processes of birds. While mere exposure to odors can alter responsiveness in some instances, even in humans (Balogh and Porter, 1986; Romantshik et al., 2007), there are numerous sources of reinforcement present soon after birth. For example, odor preferences are reinforced in the suckling environment by the warmth of the mother, tactile stimulation from licking, and the milk. Simply pairing an artificial odor with a warm surface or with tactile stroking is sufficient to establish conditioned olfactory preferences in rat pups (Alberts and Brunjes, 1978; Alberts and May, 1984; Dominguez et al., 1999). In general, neonates detect and find attractive the odorous components of amniotic fluid, particularly those of their own mothers (Schaal et al., 1998; Teicher and Blass, 1977; Hepper, 1987), likely reflecting intrauterine experience and possibly explaining their attraction to nipple-related odors and to other maternal secretions around the time of birth (Schaal et al., 1994). Such attraction aids in guidance to the nipple and alters their general motor activity and arousal (for review, see Porter and Schaal, 2003). Parturient females of many species engage in self-grooming that deposits saliva and amniotic fluid on their ventral and nipple regions, and these secretions can carry, in some instances, the chemical message that directs the first suckling episode of the newborn (Teicher and Blass, 1976, 1977). Even human infants, who preferentially exhibit head orientations toward maternal breast odors within the first few minutes of life, are influenced by prior experience with amniotic fluid. As noted by Porter and Winberg (1999), "The role of maternal olfactory signals in the mediation of early breast-feeding is functionally analogous to that of nipple-search pheromone as described in nonhuman mammals. To some extent, the chemical profile of breast secretions overlaps with that of amniotic fluid. Therefore, early postnatal attraction to odors associated with the nipple/areola may reflect prenatal exposure and familiarization" (p. 439). Some foods ingested by the mother markedly influence the smell of the amniotic fluid and an infant's attraction to it (Mennella et al., 1995), as well as affect the flavor of the mother's milk.

It is of interest that, in rat pups, pairing an odor with an aversive shock prior to 10 days of age leads to an odor preference, whereas such pairing after 10 days leads to an odor aversion (Moriceau and Sullivan, 2004). The latter can be mitigated by having the mother present, suggesting the coexistence of "attachment" and "adult-like" neural systems after the age of 10 days (Shionoya et al., 2007; Moriceau and Sullivan, 2006).[1] The imperviousness to aversive odor conditioning before the age of 10 days presumably relates to the need to have a stable mother-infant bond during a period when the pup may have to be roughly handled and independence from the mother is not viable.[2]

A number of cross-fostering studies find it is the odor of the species or subspecies of the cross-fostered parent, not that of the genetic parent, which largely establishes subsequent social and mating preferences. In some cases, male odor is important. For example, estrous *Mus musculus domesticus* female house mice reared by both parents prefer, after weaning, the odors of *M. musculus domesticus* to those of *M. musculus bactrianus,* whereas analogous females reared only by their mothers, in the absence of adult males, show no such preference (Mainardi, 1963). Female house mice cross-fostered to pigmy mice (*Baiomys taylori*) prefer the odor of the pigmy mice to that of house mice in adulthood (Quadagno and Banks, 1970). Male white-footed mice (*Peromyscus leucopus*), cross-fostered to grasshopper mouse dams (*Onychomys torridus*), switch their social preferences to the grasshopper mice (McCarty and Southwick, 1977).[3] Interestingly, gerbils (*Meriones unguiculatus*) reared with parents whose midventral sebaceous glands were surgically removed show lower preferences for such odors in adulthood and engage in less social behavior with opposite-sexed conspecifics than gerbils raised with parents having such glands (Blum et al., 1975). Similarly, female house mice reared with mothers whose preputial glands have been removed prefer to be with house mice without preputial glands in adulthood (Hayashi, 1979).[4]

Learned responsiveness of rodents to odorants in the suckling environment or nest is not confined to natural, biological, or animal odors (Cornwell, 1976; Gregory and Bishop, 1975; Galef and Kaner, 1980; Janus, 1993; Macrides et al., 1984). Any of a number of artificial odorants (e.g., ethyl benzoate, acetophenone, methyl salicylate, citral, cinnamon, cumin, and various perfumes and colognes, such as Parma Violet perfume and Yardley's Red Roses cologne) placed in the rearing environment can take on the same meaning as natural biological stimuli and alter subsequent preferences for scented situations or scented conspecifics in later life (Carter, 1972; Carter and Marr, 1970; Janus, 1989, 1993; Fillion and Blass, 1986a). For example, rats reared on lemon-scented bedding from birth to weaning acquire a seemingly permanent preference for nesting in lemon-scented surroundings (Rodriguez Echandia et al., 1982).[5] Adult rats previously reared with mothers and littermates odorized by artificial odors prefer conspecifics odorized with such odors and are less responsive sexually to unodorized conspecifics (Marr and Gardner, 1965; Marr and Lilliston, 1969; Fillion and Blass, 1986b). The same is true for mice, as shown by Mainardi et al. (1965). These investigators reared male and female house mice (SWM/Mai strain) with perfumed or non-perfumed parents until the age of 21 days, when they were weaned and separated into like-sex groups. When tested in estrus at 8 months of age, 48.2% of the perfume-reared females spent more than 60% of their time with a perfumed male, compared to 21.4% of the normally reared females; 67.8% of the normally reared females spent more than 60% of the test time with the non-scented males, compared to 27.6% of the perfume-reared females.

It should be emphasized that neonatal learning is also not confined to olfaction and similar information is often also gleaned from other sensory cues, emphasizing the multimodal mediation of early sensory experiences (Beach and Jaynes, 1954). For example, unlike their normally reared counterparts, male sheep and goats cross-fostered to the opposite species show a nearly exclusive preference for faces of females of their foster species. Cross-fostered females also exhibit, relative to normals, an increased preference for the faces of the cross-fostered species, although their preferences are more or less equally divided among the faces of the genetic and cross-fostered species (Kendrick et al., 2001). Kendrick et al. (2001) pointed out the important role of experience in this process: "These results provide strong evidence that social and sexual preferences are primarily determined by maternal and social rather than genetic influences even in mammals and that the effects are stronger and more durable in males than in females" (p. 334).

The aforementioned findings largely compromise the tenets of the pheromone concept when applied to mammals and suggest the olfactory system has evolved, like the visual system, to recognize, remember, and prefer variant meaningful salient features of the general environment. Rather than being automatons, most mammals respond to the vast majority of environmental chemicals on the basis of experience. As noted by Helen Keller's statement at the beginning of this chapter, even our own species readily learns to identify odors on the basis of their relationship to their physical sources, as denoted by their names (e.g., rose, candy, pizza, gasoline, fish, licorice, chocolate, leather, mint, seashore, perfume, etc.).

Olfactory Learning in Later Life

Olfactory learning analogous to that which occurs in the litter situation or soon after weaning also occurs in adulthood. Most adult mammals have the ability to rapidly acquire and maintain memories for many types of odors, not just natural social ones. For example, mice (*Mus musculus*), rats (*Rattus norvegicus*), hamsters (*Mesocricetus auratus*), Belding's ground squirrels (*Spermophilus beldingi*), and guinea pigs (*Cavius porcellus*), species tested extensively on this point, are capable of remembering hundreds of odors encountered in adulthood (Johnston, 1993; Beauchamp and Wellington, 1984; Mateo and Johnston, 2000; Brown, 1988; Mossman and Drickamer, 1996; Larson and Sieprawska, 2002). This is possible in some cases even after having encountered the odor on a single brief occasion. In some species, social interactions aid in establishing such recognition (Todrank et al., 1999). Pairing an artificial odor with sugar establishes strong preferences for that odor over non-reinforced odors in mice for as long of periods as have been tested (e.g., two months) (Schellinck et al., 2001). Interestingly, rats can reportedly distinguish among individual human scents without training (Krutova and Zinkevich, 1999).[6]

Contrary to the tenets of the pheromone concept, experience with odors later in life can significantly influence subsequent odor-mediated behavioral and sexual preferences.[7] For example, adult social experience with C57 male mice can reverse the social preferences of SEC1Re/J mice from their own strain to that of the C57 strain (Albonetti and D'udine,

1986). Rats, dogs, and a number of other mammals develop preferences, or markedly increase preexisting subtle preferences, for estrous over diestrous female odor as a result of adult sexual experience (see Chapter 5). In one study artificial odors (i.e., musk oil, oil of clove, lemon/lime, and cherry) were applied every other day on four occasions *after weaning* to four groups of spiny mice (*Acomys cahirinjus*), each composed of two siblings and two non-siblings (Porter et al., 1983). Each group received a single odor. Later preference tests conducted with odorized kin and non-kin strangers found that the mice interacted, for all practical purposes, only with mice having the same odor to which they had been exposed as adolescents, regardless of genetic relatedness. This work shows that any salient odor, not just putative "pheromones," can have a significant influence on social behavior in adulthood.[8]

The preference of female house mice for male mouse odors is learned. Moncho-Bogani et al. found that female mice, which have been sequestered from mature male mouse odors from an early age, show no preference for male over female mouse or castrated male mouse odors until they have had direct physical contact with male scents, presented in soiled bedding, via the vomeronasal organ (Moncho-Bogani et al., 2002, 2004, 2005). Once such contact has occurred with bedding from multiple males, they exhibit a generalized preference for male mouse over female mouse odor, presumably via associative conditioning. These authors suggest that an innate preference exists for male over female stimuli when the vomeronasal system is stimulated, whereas this is not the case for the rapid-learning main olfactory system, and that vomeronasal sensing of such chemicals is intrinsically rewarding. As noted earlier, when urine from individual males, rather than grouped males, serves as the stimulus, the conditioned preference mediated by the main olfactory system is specific to the urine of the male donor, not to urine from different males, implying considerable specificity in what is learned (Ramm et al., 2008).

In a profound observation antithetical to the pheromone concept, Johnston and Jernigan present evidence that hamsters develop "integrated multiodor representations of other individual hamsters" via their experience with them, implying a "higher-order processing system that categorizes stimuli according to their significance and not strictly by their sensory characteristics" (Johnston and Jernigan, 1994, p. 133). These investigators show that male hamsters come to recognize, as a result of brief social encounters, the concept of an individual female from multiple odor sources (e.g., flank gland and vaginal secretions) in a manner conceivably analogous to our ability to recognize a given friend from his or her voice or picture. Using an odor habituation paradigm, *vaginal secretions* from a familiar female hamster were repeatedly presented to male hamsters in their home cages for 5-minute periods separated by 15-minute intervals. Over the course of a few trials, the males gradually decreased their sniffing and exploratory behavior directed toward the secretions, reflecting a loss of interest. When the *flank gland secretions* of this female were presented along with those of another familiar female, the males exhibited marked interest only in the flank gland secretions of the other female. This suggests that the males habituated to a generalized olfactory concept of the original female, not just to her vaginal secretion odors. The authors believe that this phenomenon is unlikely due to similarities in vaginal and flank gland chemicals, since hamster vaginal secretions are pungent, odorous, and comprised of many small volatile molecules, a number of sulfur compounds, and proteins, whereas hamster flank gland secretions are generally weak in odor, being largely composed of sebaceous lipids and associated breakdown products (Johnston and Jernigan, 1994). Prior physical contact is needed between the male and the original female to produce this effect, such as that achieved by the male touching his nose to the female's flank gland through a screen (Johnston and Peng, 2008), suggesting such recognition may involve the vomeronasal organ.

Hormone levels, as well as sexual preferences, can be influenced by adult experiences with artificial odors, stressing the point that "pheromones" are not unique in this regard. For example, male rats exhibit an increase in testosterone and luteinizing hormone following exposure to the wintergreen-smelling odorant, methyl salicylate, when the odor had been paired with previous copulation (Graham and Desjardins, 1980). Males mated with females scented with almond extract ejaculate in subsequent tests with almond-scented females first and more frequently than with non-scented females, a phenomenon related to the degree of satiety and the rat's motivational state (Kippin and Pfaus, 2001a). Such rats tend to preferentially mount the almond-scented females immediately prior to ejaculation, implying that the preference may be conditioned to the ejaculatory event per se, although the presence of the female during the post-ejaculatory period seems necessary for the effect (Kippin and Pfaus,

2001a, 2001b). Female rats also develop olfactory-conditioned partner preferences for cues associated with sexual rewards, including cues related to copulation and cues related to ejaculation (Coria-Avila et al., 2005).

The influence of odors learned in adulthood extends to the social communication of what foods are safe to eat, implying considerable complexity in information transfer. For example, rats alter their preferences for foods eaten by other rats with which they socialize, a phenomenon that is accentuated in protein-deprived rats (Galef et al., 1991; Beck and Galef, 1989). In one study it was found that rats that encounter other rats that have eaten banana-flavored food pellets are more likely to enter a T-maze arm known to lead to such pellets (Galef et al., 1997). In another study it was shown that a rat will exhibit an enhanced preference for a cinnamon- or cocoa-flavored food recently eaten by a healthy rat placed in its cage for a brief period, a preference that does not generalize to similarly scented nest materials or nest boxes (Galef et al., 1994). Rats made sick after eating a series of novel foods in succession are less likely to exhibit a conditioned aversion to those foods whose odors were previously experienced on the breath of healthy conspecifics to which they were briefly exposed (Galef, 1986). This is not simply a function of greater familiarity with the novel food odor. Rats that have learned an aversion to a flavored fluid that are allowed to briefly interact with healthy rats that have drunk that fluid without adverse consequence increase their intake of the averted fluid relative to controls that have had no such social interaction (Galef et al., 1997). Conversely, rats that have consumed a novel substance and then are exposed to a sick rat develop an aversion, albeit weak, to the novel substance, a phenomenon termed the *poisoned-partner effect* (Batsell et al., 1999).

Even after the meaning of an odor, including its health consequence, becomes manifest to an animal, its behavior is not reflexive or invariant toward that specific stimulus, as implied by the pheromone concept. For example, a male mouse typically responds to the scent marks of another male by depositing urine on or around such marks, as do a number of other mammals (see the following chapter). However, if the mouse ends up on the losing end of a fight with the other mouse, it will become subordinate to it and will cease such scent marking even when it is housed in close proximity to the marked area, its circulating testosterone is main-

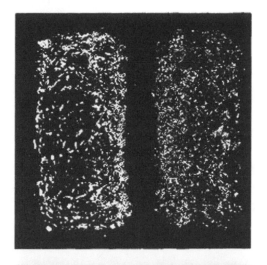

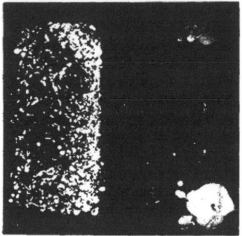

FIGURE 3.1. *Upper photo:* Ultraviolet photographs of overnight marking patterns of socially naïve, male, wild house mice, one in each compartment of a 12 × 12 inch cage, separated from each other by a wire mesh partition. *Bottom photo:* Marking patterns following establishment of dominance in an agonistic encounter, with marking of dominant animal shown on left and subordinate shown on right. From Bronson (1976a); used with permission, copyright Elsevier 1976.

tained at a high level by a silastic implant, and the other male is not present (Maruniak et al., 1977) (Figure 3.1). Hence, in this case the mouse's scent marking is not an invariant response to a pheromone or even to endogenous levels of testosterone (which are usually correlated with scent-marking behavior and glandular secretions; Beauchamp, 1974; Mykytowycz and Dudzinski, 1966), as learning has intervened (see the "male mouse aggression pheromone" in Chapter 5). Social and contextual factors also influence responses to purported pheromones. For example, dodecyl propionate, a putative maternal pheromone isolated from rat preputial glands, is said to attract adult rats and to play a key role in regulating postpartum maternal licking of the anogenital area of pups, a behavior critical for initiation of defecation (Brouette-Lahlou et al., 1992). When placed on rice, however, dodecyl propionate deters licking and ingestive responses (Arnould et al., 1994). Female mouse odor preferences are also influenced by social influences. Thus, female mice from social groups with neighboring social groups show a stronger relative preference for the scent marks of dominant males from their own groups than do females from groups with no neighbors (Heise and Hurst, 1994).

The examples outlined in this chapter clearly demonstrate that the meaning of many chemical stimuli commonly termed *pheromones* is learned and context dependent, and that in light of a dynamic perceptual system one cannot assume that most responses to conspecific biological secretions are reflexive or innate. Attempts to distinguish between inherent and learned responses are very difficult to make, particularly since postpartum responses can be tempered by prior intrauterine learning. Learning is a particularly salient feature of mammals, largely explaining why searches for simple underlying invariant stimuli generally have been snark hunts. Chemically mediated behaviors or responses cannot be assumed to be more primitive or less influenced by learning than behaviors or responses mediated by nonchemical stimuli, for which such terms as *visuomones, audiomones,* and *touchomones* have not been employed. Even if it some natural or biological odors are inherently more attractive than others, describing such odors as "pheromones" does not seem to serve any meaningful scientific end.

FOUR

Scent Marking

···

After the error of atheism, there is nothing that leads weak minds further
astray from the paths of virtue than the idea that the minds of other
animals resemble our own, and that therefore we have no greater right
to future life than have gnats and ants.

René Descartes, *Passions of the Soul,* 1649

In this chapter the influences of learning will be shown to be particularly salient in responses to conspecific scent marks or depositions commonly said to contain pheromones, emphasizing the fact that invariant responses to chemical stimuli are not the norm in mammals. Scent marking is one of the most common socially related behaviors of mammals. As a sexually dimorphic trait, species-specific behaviors used in scent marking, such as the leg lift urination posture of the dog, depend in large part upon intrauterine exposure of the brain to testosterone, sensitizing neural structures to activation by testosterone in adulthood (Beach, 1974) (Figure 4.1). Scent marking is present within at least 15 of the 18 mammalian orders and within the entire range of social structures (e.g., pair-bonded territories; single-male territories, or demes, with clearly defined boundaries; single-male territories around harems, packs, etc.) and is a critical means by which many macrosmatic mammals provide information regarding their location and physiological status. In most species, males do the majority of the marking and, when scent glands are involved, have larger scent glands than females (Figure 4.2). This behavioral and structural sexual dimorphism appears analogous to that of many birds. Thus, just as the males of most avian species are brightly colored and are more involved in territorial and courtship displays than

FIGURE 4.1   Twelve elimination postures observed in purebred beagles. In 60 males and 53 females tested, males predominantly urinated using the elevate posture (97%) and the raise posture (2.1%). Females used a wider range of postures, including the squat (68%), the squat raise (19.3%), the raise (4.6%), and the flex-raise (3.1%). Modified from Sprague and Anisko (1973); used with permission of Koninklijke BRILL NV.

the females, most male mammals are more brightly odored, if you will, and more involved in territorial scent-marking activities (Doty, 1974).

Like the individual plumage of birds or the characteristics of human faces, the scent sources of mammals are complex. The components of scent glands and urine, for example, can number in the hundreds and even thousands, depending upon such non-mutually exclusive factors as genetics, commensal bacteria, season, habitat, diet, housing conditions, and endocrine state (Schellinck and Brown, 1999; Pohorecky et al., 2008). In the case of urine, Rasmussen and Krishnamurthy (2000) state:

> The number and complexity of chemical compounds in urine is enormous—up to several thousand different compounds. Large volumes of emitted urine may remain in liquid form for several days. Thus, urine represents a relatively persistent signal. In addition, compounds or blends impart dif-

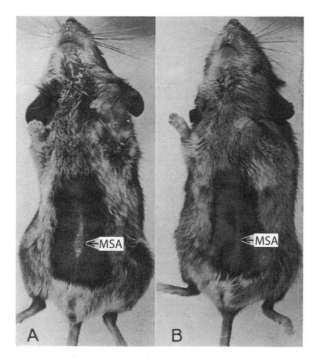

FIGURE 4.2. A. Shaven belly of male deer mouse (*Peromyscus maniculatus bardaii*) exposing the large midventral sebaceous scent gland area (MSA). B. Same area of female deer mouse showing much smaller scent gland region. Modified from Doty and Kart (1972).

ferent messages, and various compounds degrade at different rates, pro- viding a temporal indication of signal age and thus the presence of the emitter... The presence of proteins may retard this degradation or affect the release rate of smaller, more volatile compounds. (p. 408)

The influence of diet on the attractiveness of urine or scent marks has been demonstrated in a range of species, including guinea pigs (Beauchamp, 1976), meadow voles (Grigoriadis et al., 1989), and beavers (Tang et al., 1995). For example, meadow voles are more strongly attracted to scent marks of conspecifics having higher-protein diets (Grigoriadis et al., 1989), and many of the large number of phenolics and terpines found in the castoreum of beavers are diet-derived, conceivably advertising to potential mates the nutritional status of the individual and, indirectly, the food supply in his territory (Tang et al., 1995). Dietary factors can mask individual odors that may have a strong genetic determinant (Shellinck et al., 1997) and microbial products often produce the characteristic odors of scent gland secretions. Like the situation with complex odor mixtures discerned by humans, the odorous elements of scent marks presumably are synthesized into relatively uniform percepts reflecting the nature of neural processing within the olfactory bulb and other brain structures (Doty and Laing, 2003).

The complexity of scent marking is clear when one examines the various functions that such marks serve (for reviews, see Ewer, 1968; Brown and MacDonald, 1985; Doty, 1986b). In some cases, scent marking is involved in intrasexual competition, dominance assertion, and the attraction of mates. In many cases, scent marking imparts an individual or group odor derived from the dominant individuals (Mykytowycz, 1973; Schultze-Westrum, 1969). Such marking familiarizes members of the group with the odor of the individual or individuals doing the marking, leading to a higher probability of the marker's acceptance in mating and a reduction in the cost of agonistic encounters between competitors and dominant resource-holding animals. Importantly, scent glands on different regions of the same animal may serve different functions. For example, glands on the inside of the back leg of black-tailed deer, termed *tarsal glands,* seem to be mostly involved in sex, age, and individual recognition, whereas glands on the outside of the back leg, termed *metatarsal glands,* aid in communicating emotional state, such as fear. Glands on the forehead are frequently used in territorial marking (Müller-Schwarze, 1971) (Figure 4.3).

Some scent marks or scent-marking behaviors seem to influence, if you will, the mood, motivation, or confidence of both the marker and other conspecifics. Thus, when two rabbits are brought together in a pen to which the experimenter has added fecal pellets from only one of the pair, the rabbit whose odors have been added will more than likely be the victor in the ensuing agonistic encounter (Mykytowycz, 1973). The same phenomenon occurs when the feces of a familiar opposite-sexed partner are substituted for the individual's own feces. Aside from marking areas where other conspecifics have marked, foxes, coyotes, and wolves urine mark on sites where food odor persists but limited food is present. A forager thus saves time by not investigating such sites (so-called bookkeeping) (Henry, 1977; Harrington, 1981).

While some proponents of the pheromone concept infer that the con-

FIGURE 4.3. Territorial male Thomson's gazelle (*Gazella thomsoni*) marking vegetation with his preorbital gland. Photograph courtesy of Ferrero-Labat (www.ardea.com).

stituents of scent glands (e.g., musks) serve as pheromones that, in effect, drive conspecific males away from established territories, as occurs in some insects (Wilson, 1972), neither the responses of mammals to scent gland constituents nor the hormonal control of scent marking are simple or reflexive. In fact, scent marks are hardly ever completely avoided by animals that have not previously encountered the depositor of the mark, and only rarely repel intruders from regions that have been marked (Fullenkamp et al., 1985; Gosling, 1982; Price, 1975; Richards and Stevens, 1974; Jones and Nowell, 1973; Mucignat-Caretta, 2002).[1] If anything, conspecific scent marks attract conspecifics and lead to increased counter-scenting activities (Baran, 1973; Gosling et al., 1996b), although this can be modulated by a number of factors, including the novelty of a conspecific odor and the presence of a predator's odor near a conspecific's mark. The latter implies that there is a tradeoff between territorial defense and predator avoidance (Rosell and Sanda, 2006). Importantly, a number of nonterritorial species scent mark and some novel odors—including artificial odors such as amyl acetate—can elicit scent-marking behaviors at the same frequency as more natural odors, at least in some animals (Hopp and Timberlake, 1983).[2] In a variety of canids, scent rubbing seems to be elicited by novel stimuli, being most prominent for (1) scents of a class not usually encountered in the animal's environment (e.g., artificial odors), (2) familiar scents whose character deviates from the norm, and (3) scents that otherwise induce a strong aversion or attraction (Ryon et al., 1986).

Gosling, using data largely derived from studies of deer and other ungulates, presents a theory of scent marking based upon the general principle that scent marks are not reflexively avoided (Gosling, 1982). In addition to containing intrinsic (e.g., mark density and intensity) and learned (e.g., memorization of past opponents) information, Gosling's theory focuses on the notion that scent marking allows for the scent matching of new opponents with the smell of recently encountered marks. This essential element of his theory, grounded in contemporary evolutionary thinking, is as follows:

> The function of territorial marking is to provide an olfactory association between the resident and the defended area which allows intruders to identify the resident when they meet and thus reduce the frequency of escalated agonistic encounters. An animal that can defend an area long enough to mark it comprehensively is likely to win most encounters with intruders because of its physical quality (intruders will vary in quality). It [the resident] is also more likely to escalate to overt fighting since, with a more detailed knowledge than an intruder of the territory's resources, it has more to gain by retaining ownership (Dawkins and Krebs, 1978). It will thus pay low status intruders to withdraw from encounters with an identified resident. Only a minority of high status intruders might choose to escalate an encounter with a resident in an attempt to displace it. Marks thus provide a way for an intruder to assess the quality of a potential

competitor so that they can avoid escalation in encounters that have a high risk of injury, except when the potential benefits of an escalated encounter are also high. The advantage to the resident is that by providing these means of assessment it avoids the costs of establishing dominance by threat or overt aggression towards *every* intruder.

The suggested mechanism of assessment is that intruders compare the scent of any animals they meet with the memorized scent of marks that they have encountered in the vicinity. When these scents *match,* then the resident is identified and the intruder responds appropriately, usually by withdrawal. This simple physiological mechanism provides a precise means of competitor assessment in the territorial context and is central to the hypothesis. (p. 94)

Gosling goes on to point out that the number of marks encountered by an intruder may provide important additional information, including the duration and/or frequency of residence and the dominance status of the resident (given the relationship among marking frequency, dominance, and testosterone titer). In effect, marks signal potential fitness costs to opponents in general rather than just to intruders into a territory (Gosling, 1990). Gosling and Roberts (2001) state:

If the signaler is absent when the receiver detects a mark, these costs will depend on the probability that the signaler will return, its relative competitive ability, and the relative value of any marked resources to the two individuals. The involvement of these various factors may explain why the responses to intruders to scent marks are so variable. Some intruders into territories appear to be undeterred by scent marks. They smell the marks but then move on through the territory or stay to use its resources. In other cases, best known from studies of mice, males avoid scent-marked substrates, especially when they are of low competitive ability (Gosling et al., 1996a; 1996b) or when the scent is from dominant males (Jones and Nowell, 1974b; Hurst, 1993), and are more reluctant to risk or prolong fights with males whose scent suggests that they are territory owners (Gosling and McKay, 1990; Hurst et al., 1994). (p. 173)

A number of phenomena are explained or predicted by Gosling's theory that are not well explained or predicted by other theories of scent marking, including those suggesting scent marks contain pheromones (e.g., why some animals anoint themselves with urine, why some animals bathe in dung or other smelly substances, why scent marking in some species is a ritualized component of agonistic encounters, and why residents allow intruders to smell them). Predictions from his theory, which are supported by numerous field and laboratory studies, include the following: (1) the owner should mark the territory in a way that maximizes the chance that marks will be detected by an intruder; (2) the owner should mark itself with the substances used to mark the territory except when the odor is available to an intruder at its site of production or in another available substance; (3) the owner should make itself available for scent matching by the intruder; (4) the owner should remove or replace marks in the territory that do not match its own odor; (5) intruders should seek out or otherwise detect the characteristic scent marks of territory owners; (6) when intruders meet animals that could be the territory owner they should smell, and perhaps taste, any secretion or odor used in marking the territory; (7) low-status intruders should withdraw if the scent of a possible territory owner matches that of marks detected previously; and (8) high-status intruders should usually withdraw when the scent of a possible territory owner matches that of marks detected previously. As noted by Gosling and others, such predictions are supported in a wide range of vertebrates, including salamanders and rodents (see, e.g., Luque-Larena et al., 2001; Simons et al., 1994). In effect, scent gland secretions, as well as urine, feces, saliva, and other excretions, are akin to traffic sign posts or pictures, imparting significant information about the environment, including the whereabouts, age, gender, social status, and emotional state of the involved individuals.

It is noteworthy that a number of mammals can recognize not only the most recently deposited scent on substrate previously marked by another conspecific, but the scent over which the last mark was deposited (for reviews, see Johnston, 1999; Ferkin, 1999). Laboratory studies of hamsters and meadow voles suggest, however, that the upper-most mark is of special significance, seemingly being tagged in memory as being more important than the lower-most mark. This phenomenon is apparently independent of the mark's age or freshness (Ferkin, 1999), although age can be used in other contexts and, in the case of mice, may be a determining factor (Rich and Hurst, 1999). Dogs can discern the direction of a human scent trail by differences in age of the footprints along the trail

(Wells and Hepper, 2003) and estrous rat urine loses its attractiveness over a relatively short period of time after exposure to air (Lydell and Doty, 1972). Interestingly, hamsters are able to analyze the spatial configuration of two scent marks (e.g., an interrupted versus continuous scent mark) for information about which individual marked over the other (Johnston and Bhorade, 1998; Johnston, 1999).

In light of such complexities, which vary among species, the application of the pheromone concept to "territorial" behaviors related to scent marking seems overly simplified, leading to conclusions that are not supported by the data and underestimating the nature of the biological and psychological processes involved. The complex nature of scent marking and its associated information transfer suggests that scents for many mammals may reflect cognitive processes analogous to visual or auditory signifiers employed in human language.

The Elusive Snarks

Case Studies of Nonhuman Mammalian "Releasing" Pheromones

....

We have sailed many months, we have sailed many weeks,
(Four weeks to the month you may mark),
But never as yet ('tis your Captain who speaks)
Have we caught the least glimpse of a Snark!
    From the Bellman's Speech, in Lewis Carroll's *The Hunting of the Snark*

B y this time some readers may have begun to entertain the thought
that mammalian pheromones may truly be snarks, being difficult
if not impossible to identify. As indicated in Chapters 2 and 3, hundreds
of studies have appeared in the past half century claiming that the mammalian behaviors or endocrine events they describe are due to pheromones. However, as indicated by Buck (2000), "few mammalian pheromones have been identified" (p. 611). I submit that such agents do not
generally exist.

This chapter is a critical analysis of the major reports of identification
of mammalian "releasing" pheromones, as well as a number of reports
claiming the existence of yet-to-be-identified pheromones. A similar analysis of "priming" pheromones is made in Chapter 6. As will become apparent, single chemicals or small sets of chemicals that accurately mimic
the effects of whole secretions are generally nonexistent and most bioassays have employed sexually and socially experienced animals as subjects, confounding the bioassay with learned responses. In some cases,
less than half of the subjects respond in the expected manner and in a
number of cases the findings have failed to be replicated.

In a series of highly publicized studies in the late 1960s and early 1970s, vaginal secretions from female rhesus monkeys were reported to contain pheromones that elicit copulatory behavior from males (Michael and Keverne, 1968, 1970a). Specifically, male monkeys were found to perform a bar-press behavior to gain access to and to copulate with estrogen-treated females on the basis of olfactory cues arising from the vagina. In related studies, ether or water extracts of vaginal secretions from estrogen-treated ovariectomized monkeys were applied to the genital area of untreated ovariectomized females. During subsequent tests, application of the extracts resulted in an "immediate and marked stimulation of the sexual activity" of the male subjects (Keverne and Michael, 1971, p. 313). The investigators concluded, "male sex-attractant pheromones, with powerful behavioral effects, are present in ether extracts of estrogen-stimulated vaginal secretions" (p. 313).

Subsequent chemical analysis of the estrous vaginal secretions resulted in the isolation of a series of volatile short-chain aliphatic acids—acetic, propionic, isobutyric, n-butryic, and isovaleric—said to be the ingredients of the active pheromonal substance (Bonsall and Michael, 1971; Michael and Keverne, 1970a; Michael et al., 1971). A mixture of these agents in specific proportions—a concoction termed *copulin*—was prepared, found to be active, and patented in several countries for employment in human perfumes, suggesting from the onset that the putative pheromone was assumed not to be species-specific. However, copulin's effectiveness in altering human sexual behavior was later found to be nil, negating the notion that the mixture of aliphatic acids had cross-species effects that generalized to humans (Morris and Udry, 1978; Cowley and Brooksbank, 1991).

Aside from the issue of species specificity, the validity and generalizability of these findings even among rhesus monkeys was called into question in a series of studies by Goldfoot et al. (1976). These authors carefully examined the original rhesus monkey pheromone data, noting that the male responsiveness varied considerably from male to male and in some cases depended upon a particular female, irrespective of an odor cue. They pointed out that, in the seminal work (Michael and Keverne, 1970a), baseline tests consisting of as many as 60 pretests over 80 days were conducted in only 2 male subjects, and suggested that the application of the odorants after extinction of mounting could be explained on the basis of disinhibition, resulting in a resumption of mounting. Subsequently it was revealed by one of the initial copulin investigators that individual differences and social conditions did, in fact, influence the effectiveness of copulin even within the small number of subjects employed, and that some of the original females treated with "active" lavages had failed to stimulate sexual behavior in their male partners (Keverne, 1974). Particularly damning to the rhesus monkey vaginal pheromone concept was evidence that anosmia has no influence on male rhesus monkey mating behavior, implying—under the assumption made at the time that olfaction is the mediating modality—that pheromones are neither necessary nor sufficient for such behavior (Michael and Keverne, 1968; Goldfoot et al., 1978). Rhesus monkeys do not have a functional vomeronasal organ (Evans, 2006).

In contrast to the copulin studies, those by Goldfoot et al. employed relatively large numbers of subjects (Goldfoot et al., 1976). In their behavioral research a total of 19 adult male and 27 adult female rhesus monkeys were used. Unlike the copulin work, the donor females had not been recently paired with males in most of the test situations and, thus, their vaginal secretions were not contaminated by male ejaculate. Despite careful quantitative assessment of a range of sexual behaviors (approach, genital inspection, contact, mount, intromission, ejaculation) under a variety of estrogen regimens and behavioral test conditions, no statistically significant differences between vaginal lavage and control treatments could be found, save a slight tendency for ejaculate-contaminated secretions to increase some behaviors. Importantly, the amount and relative proportions of aliphatic acids found in the vaginal secretions differed markedly from those reported for copulin. For example, isovaleric acid was not detected in the secretions even after 29 days of estrogen treatment. Goldfoot et al. concluded that "comparison of our results to those from other laboratories suggests that the mechanism involved in positive effects may depend upon associative learning or upon extinction or disinhibition of sexual interest" (p. 2). Goldfoot et al.'s analytical data were in accord with a body of evidence suggesting that vaginal aliphatic acids primarily appear during the luteal phase of the cycle, rather than during the time of optimal fertility. In humans, for example, aliphatic acids are largely dependent on bacterial fermentation of glycogen, which is highest not during the time of optimal fertility, but during the luteal phase (Gregoire et al. 1973).

No other studies have expanded on this research in the ensuing three and a half decades. Despite this, reviews continue to cite the aforementioned rhesus monkey studies as strong evidence for the existence of pheromones in primates (e.g., Bakker, 2003; Cutler and Genovese, 2002).

Boar Pheromones That Stimulate Sexual Receptivity (Lordosis) in Sows

While chemically assaying androgens in pigs, Prelog and Ruzicka (1944) discovered the presence of high levels of 16-androstenes in the testes of boars. 5-α-androst-16-en-3-α-ol, commonly termed *androstenol,* was found to have a musk-like odor, whereas 5α-androst-16-en-3-one, or simply *androstenone,* had a urine-like smell (Prelog et al., 1945). These and related agents, whose origin is mainly the testes, accumulate in a number of bodily tissues and secretions of the boar, including blood, sweat, urine, saliva, and sebaceous products. These materials were implicated in the production of the "boar-taint" odor of cooked boar meat, as well as the production of odors with conspecific biological function.

Steroid-dependent odors emanating from the boar's preputial diverticulum (comprised of residual urine and secretions from the reproductive tract) became of interest to behaviorists in the 1960s because, prior to copulation, sows often nuzzle this region of the boar, implying that "odours emanating from the preputial sac might be acting as sex attractants to the female" (Booth, 1980, p. 291). Additional interest in a possible communicative role of salivary odors came from the observation that the boar, during sexual or agonistic encounters, "champs its jaws and drools a viscous, glycoprotein-rich frothy saliva which derives from its submaxillary salivary glands and which is quite unlike the more watery parotid/sublingual saliva involved in feeding" (Albone, 1984, p. 238). Individually reared boars have lower plasma levels of androstenone than group-reared boars (Narendran et al., 1980). Higher androstenone levels are found in the more dominant individuals (Giersing et al., 2000).

Claims that androstenone and androstenol are pheromones arose largely from the discovery that these chemicals, as well as secretory products from which they arise, increased the incidence of the standing response of sows when sprayed near their noses (Melrose et al., 1971; Stefanczyk-Krzymowska et al., 2000; Hafez and Signoret, 1969; Patterson, 1966, 1968). As noted by Claus and Hoppen (1979), "For female pigs in oestrus... [androstenone] is a very desirable 'male perfume' which is released by the boar's saliva before mating and stimulates the female's 'standing reflex,' thus acting as an aphrodisiac pheromone" (p. 1674). Unfortunately, some authors have mistakenly concluded that boar odor is *necessary* for lordosis in the sow. For example, Cutler and Genovese-Stone (1998) state, "In 1961 boar odor was shown to be necessary to elicit the mating stance in the female pig" (p. 355). As described in Chapter 7, a number of workers have suggested that androstenone and androstenol are human pheromones, since they are found in some human bodily secretions.

In fact, these or related agents, including full-strength preputial fluid and urine, rarely elicit the standing response in the majority of estrous sows, as shown in Table 5.1. Moreover, combinations of these agents do not necessarily increase the efficacy over that observed by the individual steroids alone, and studies in this area have generally lacked placebo controls.

Such "pheromones" are not effective in eliciting this response in diestrous sows (Signoret, 1970) and are not the sole stimuli that facilitate pressure-induced standing. Albone (1984) summarizes the complexity as follows:

> The situation is, however, a little more complicated than it seems. The oestrous sow will "stand" in response not only to olfactory signals, but also, for example, to the sound of the boar's grunting. In the natural situation the sow is exposed to a simultaneous combination of cues of many kinds, olfactory, visual, tactile and auditory, all of which play some part in stimulating the standing response, although it is clear that among these, olfactory signals are very important. Also, it is found experimentally that the oestrous sow will stand in response to the odour of boar urine or boar

TABLE 5.1. *Influence of Various Steroids Found in Sexually Mature Boar Secretions on Pressure-induced Lordosis in Estrous Sows*

Odorant	Concentration μg/ml	Sows giving a positive "back pressure" response to odorant (%)	Pigs tested (no.)
Preputial fluid and urine	Full	42	19
5 α-androst-16-en-3-one	9.12	58	50
5 α-androst-16-en-3-α-ol	4.3 53	19	
5 α-androst-16-en-3-one + 5 α-androst-16-en-3-α-ol	4.56 and 4.3 respectively	50	30
5 α-androst-16-en-3-β-ol	9.12	26	31
4,16-androstandien-3-one	9.12	53	32
5 β-androst-16-en-3-one	9.12	47	32
5 α-androstan-3-one	9.12	10	30

Sources: Based on data compiled from Melrose et al. (1971) and Reed et al. (1974); modified from Albone (1984).

preputial fluid, substances in which these particular C_{19}-$\Delta16$ steroids [androstenone and androstenol] are either absent or present at very low levels. Further, the oestrous female will respond to varying degrees to the odours of some other closely related steroids. (p. 238)

Interestingly, the lordosis facilitation by androstenone is dependent upon activation of the main, not accessory (i.e., vomeronasal), olfactory system (Dorries et al., 1997). If this agent is viewed as a pheromone, this observation is in contradiction to the argument made by some that pheromones are dependent upon the vomeronasal organ. It is noteworthy that androstenol, like a number of agents, can cross into the circulation from the highly vascularized nasal cavity to the perihypophyseal vascular complex into the arterial blood supplying the brain and hypophysis, although it is not known whether it exerts any physiological effects as a result of such absorption (Krzymowski et al., 1999; Stefanczyk-Krzymowska et al., 2000).

The most parsimonious explanation of the effects of androstenone and androstenol on lordosis in the domestic pig is that the sow is largely conditioned through experience to exhibit lordosis in response to the smell of these agents. Experience with such odors prior to adulthood might also play a role, although no research has been done on this point. It should be emphasized that sows do not lordose for all males. Thus, the female exhibits mate preferences, likely avoiding, for example, mating with littermates she remembers, as occurs in dogs (Beach and LeBoeuf, 1967). Clearly, lordosis behavior is not invariantly elicited by a "pheromone" even in pigs in which such a response can be elicited.

Hamster Vaginal Secretion Pheromones

Chemosensation plays a significant role in male hamster mating behavior. Thus, removal of sensory input via the main and accessory olfactory systems eliminates male sexual behaviors directed toward estrous females by both male and early-androgenized female hamsters (Murphy and Schneider, 1970; Doty et al. 1971; Doty and Anisko, 1973; Winans and Powers, 1974; Powers and Winans, 1973). Damaging the vomeronasal system alone irreparably reduces the arousal necessary for male mating in some, but not all, male hamsters; in others, damage to the main olfactory system alone reduces such arousal, implying variation among individuals in the relative reliance placed upon these two systems (Winans and Powers, 1977). In general, responses to distant chemical stimuli rely on the main olfactory system, whereas responses to stimuli at close quarters are subserved by both the main and accessory olfactory systems (Powers et al., 1979). Interestingly, damage to the nervus terminalis (CN O), a neural plexus located within the nasal epithelium that is generally believed unresponsive to odorants, results in a decrease in male hamster mating frequency and/or an increase in the number of intromissions required to reach ejaculation, as well as reduced attraction to conspecific vaginal odors (Wirsig and Leonard, 1987).

Its comparatively solitary lifestyle and associated dependence on chemosensory cues for mating has made the hamster the focus of many studies of putative pheromones. Female hamster vaginal secretion (FHVS) in particular has been used in attempts to isolate "sex attractant" and "mounting inducing" pheromones. FHVS is copiously produced, making it amenable to biochemical analyses, and is attractive to adult, nestling, and prepubertal hamsters of both sexes (Johnston and Coplin, 1979).[1] However, it does not invariably elicit adult mating behavior. For example, in one 18-day-long study (Johnston, 1975), FHVS intermittently rubbed on the genital regions of anestrous females led to mounting behavior by males on less than half of the trials. Without the addition of FHVS, mounting behavior occurred on more than a quarter of the trials. Furthermore, responses to FHVS may be learned. Thus, the preferences for conspecific estrous FHVS on the part of male Syrian and Turkish hamsters are significantly influenced by cross-fostering to the opposite type (Murphy, 1980). Interestingly, hamster pups "nurse" the vaginas of their mothers, a phenomenon "consistent with the hypothesis that adult sexual preferences of hamsters are influenced by exposure to their mother's FHVS" (Murphy, 1980, p. 338). Since male hamsters are equally attracted to FHVS from estrous and lactating females, one group has suggested that the odors to which male hamsters develop affinities in neonatal life may be the same as those that attract and sexually excite them in adulthood, concluding that "early experience with these maternal stimuli might be a factor in shaping the males' later sexual predispositions" (Macrides et al., 1977, pp. 384–385).

The fact that lithium chloride-based conditioned aversion to the taste and smell of FHVS is easily established in hamsters was interpreted by Johnston and Zahorik (1975) as an argument "against the common view that responses to pheromones are stereotyped and nonmodifiable." Johnston and Zahorik state, "This view is based on studies of insect pheromones and may be an inappropriate framework for investigations of mammalian systems" (p. 894).

Despite such complexities, a number of investigators have sought to identify the active pheromones contained in FHVS. To date, more than 85 volatile and many more nonvolatile compounds have been identified in FHVS, although none convincingly mimic the effects of the parent compound. Listed below are specific pheromones said to exist in FHVS.

Dimethyl Disulfide, a Sex Attractant Pheromone

Dimethyl disulfide (DMDS), isolated from *volatile fractions* of hamster vaginal secretions, was reported in one study employing gas chromatography/mass spectrometry to be a hamster sexual attractant pheromone (Singer et al., 1976). In the bioassay, *sexually experienced* male hamsters were employed, and the latency, duration, and number of animals approaching, sniffing, and digging near a section of the cage under which either DMDS, the volatile fraction, or the native vaginal secretion was located were compared. The number of animals exhibiting "positive responses" toward the whole vaginal secretion or the volatile fraction was 12 of 12. The number exhibiting such responses to DMDS varied with the DMDS concentrations employed, ranging from 42 to 67%. Of the responding subjects, the mean (± SEM) duration of approaching, sniffing and digging near the region of the cage under which the stimulus was located was 58 (±1) seconds for the whole vaginal secretion, 37 (±9) seconds for the volatile fraction, and 65 (±16), 23 (±10), 51 (±16), and 26 (±11) seconds for dimethyl disulfide at 128, 56, 22, and 2 ng concentrations, respectively. Responses to control stimuli (water) were nominal (i.e., 10 seconds or less, on average). Although the duration of behavior directed toward the 128 ng dimethyl disulfide solution was nominally longer than that directed toward the original secretion, this difference was not significant. Since a preference test with the diluent as the control odor found DMDS to be 20 to 40% less active than the whole secretion, these workers concluded that other compounds probably contribute to the attractiveness of the secretion.

Subsequently these investigators suggested that DMDS might work in concert with another chemical, methyl thiolbutyrate (MTB), which is greatly increased in amount during estrus and itself has been termed a pheromone (Singer et al., 1983). However, synergy was not observed. Thus, the mean percentage of investigation time directed toward the scented hindquarters of an anesthetized male stimulus animal by a sexually experienced male hamster was 48% for the solvent, 70% for DMDS, 63% for DMDS in combination with MTB, 65% for MTB alone, and 86% for the natural vaginal secretion. In other words, while the vaginal secretion increased the investigation time relative to the control solvent by 79%, the purported pheromones, DMDS and MTB, did so only by 46% and 35%, respectively. The combination of the latter two secretions was less effective than either secretion alone, increasing investigation time over the control by only 31%.

Other studies have stressed the fact that responses to DMDS vary across subjects and that there is reason to question whether this agent is a sex attractant at all. O'Connell et al. (1978) found that nearly half (i.e., 6 of 13) of the *sexually experienced* male hamsters they tested failed to respond to DMDS, and of those that did respond, the duration of sniffing was shorter than that directed toward the whole secretion. Petrulis and

Johnston (1995) reasoned that if DMDS is truly a sex attractant pheromone, then male hamsters should spend more time than female hamsters
investigating it and the attraction to the substance by males should be
testosterone-related. In the first of two experiments, these authors found
that *sexually experienced* males investigated female vaginal secretions
more than did females, but this was not the case with DMDS, where both
sexes equally investigated the agent. In the second experiment, *sexually
experienced* castrated males given testosterone investigated the vaginal
secretions more than castrated males not given testosterone. However,
neither castration nor testosterone repletion influenced attraction toward
DMDS or a control odor, leading these authors to conclude that "DMDS
does not elicit sex differences in attraction and that in males the attraction
to DMDS is not dependent on gonadal hormones. These results suggest
that DMDS is not a sex attractant by itself nor is it a major component
of an attractant mixture" (p. 779).

Aphrodisin, the Mounting Pheromone

Aphrodisin is a 17 kDa water-soluble glycosylated protein consisting
of 151 amino acids that has been identified within the nonvolatile high
molecular weight fractions (HMFs) of FHVS, producing, in one study
(Singer et al., 1986), intromission attempts similar in number to those
produced by the parent HMF fraction in sexually experienced male subjects. However, the number of such attempts was only about half of those
observed for the whole vaginal secretion in other studies (e.g., Clancy et
al., 1984). Proteolytic enzymes were found to decrease both the amount
of aphrodisin and its behavioral effectiveness, as did ovariectomy and
hypophysectomy (Singer et al., 1984).

To date, there is no evidence that aphrodisin interacts with chemore-
ceptors in either the main or accessory olfactory systems, and it has been
suggested that it may, in fact, serve as a transporter for smaller molecules
that may be the ligands. Aphrodisin is homologous with odorant-binding
protein, and binds a wide range of volatile ligands with similar affinity
(Briand et al., 2000). As noted by Singer (1991), "If aphrodisian does have
a [bound] ligand essential for activity, then this ligand must be tightly
bound to survive repeated dialysis and gel permeation chromatography in
aqueous buffers. The ligand itself apparently has no activity or it may

decompose when it is separated from the protein. In either case it is clear that the female hamster pheromone requires the protein, and it probably requires a ligand, for biological activity" (p. 629). Elsewhere Singer et al. (1987) indicate that "this is a question of considerable interest, because it is still widely presumed that pheromones are small molecules. No other proteinaceous pheromones are known in mammals, although they have been reported to occur in various microorganisms" (p. 294).

It would seem that hamster vaginal pheromones have still not been captured and that research in this area is dwindling. The last published paper on this topic was over a half decade ago (Jang et al., 2001) and the next most recent paper occurred in 1996 (Kroner et al., 1996).

Methyl *p*-hydroxybenzoate, the Sex Pheromone of the Dog

It is well established that *sexually experienced* male dogs, but not sexually inexperienced ones, show robust preferences for the odor of estrous females relative to diestrous ones in a broad range of behavioral tests (Beach, 1949; Doty and Dunbar, 1974; Dunbar et al., 1980). The sources of the odors that are preferred include urine and vaginal secretions, but not anal sac secretions (Doty and Dunbar, 1974).

A study published in *Science* in 1970 named methyl *p*-hydroxybenzoate as the pheromone that elicits the male's sexual interest in the estrous female (Goodwin et al., 1979). In this study, three purebred beagles, four beagle-spitz crosses, and one German shepherd served as subjects, reflecting "fully adult male dogs varying in breed, size, age, and prior sexual experience" (p. 559). Methyl *p*-hydroxybenzoate was identified by gas chromatography/mass spectrometry within unique peaks "generated from samples on the days of maximal behavioral receptivity" (p. 559) of the three female dogs. The behavioral assay consisted of pairing the 8 male and 3 female dogs in a "systematic rotating fashion" in small outdoor enclosures. The dog pairs were first placed together for 5 to 7 minutes, after which the bitch was removed and the methyl *p*-hydroxybenzoate was absorbed on the tip of a saline-soaked cotton applicator that was inserted into her vagina and rubbed over the vulvar area. She was then returned to the male. The authors found that "the results were unequivocal. In no case did the males exhibit overt sex behavior during the 5 to 7 minutes when initially paired with the female. In 18 of 21 trials conducted intermittently over a period of 6 weeks [after addition of the methyl p-hydroxybenzoate], the males attempted to mount the female after reentry... Positive responses were also elicited when the compound was applied to a spayed German Shepherd, an anestrous Dalmatian (house pet), and two other beagles not used in previous behavioral tests" (p. 560).

Similar application of 4-hydroxy-3-methoxybenzaldehyde, methyl salicylate, p-hydroxybenzoic acid, and ethyl p-hydroxybenzoate alone "caused increased ano-genital investigation" (p. 561). Mounting attempts were seen in two of six trials with propyl p-hydroxybenzoate. The authors confided that "there are many other components of the vaginal odor that we have not yet identified and tested for behavioral effects. Some of these odorants may be active alone or as synergists to methyl p-hydroxybenzoate" (p. 561).

In an attempt to replicate these findings, Kruse and Howard performed an extensive study in which a saline control was employed (Kruse and Howard, 1983). In the first of three experiments, five male and five female beagles, all of whom had mated with one another on at least one prior occasion and hence were *sexually experienced,* served as subjects. Testing consisted of (1) a baseline test period, where male-female pairs were observed when no odors were applied to the anestrous female; (2) a test period wherein a saline-soaked cotton applicator was inserted into the vagina and rubbed over the vulvar area; and (3) a test period wherein the female's vulva and vagina were swabbed with a saline-soaked cotton applicator containing methyl p-hydroxybenzoate.[2] This order of treatments was used in all trials since applying methyl p-hydroxybenzoate in either the first or second test would contaminate the subsequent tests. Eleven sets of tests with the three treatments were performed. Additionally, three baseline control periods followed by saline-only tests, and four baseline control periods followed by saline + methyl p-hydroxybenzoate tests, were employed.

Although both the saline and methyl p-hydroxybenzoate induced, in some cases, slightly more copulatory behaviors than noted during the control period, no statistically meaningful differences were noted between these two conditions. In Experiment 2, eight additional trials were performed with four males and four anestrous females, in which the cotton swabs were rubbed on the outside of the vulva and on the backs of the flanks, but not inserted into the vagina. The number of tests where mounting attempts occurred was equivalent for the saline and methyl p-hydroxybenzoate sessions, and in every case where mounting was observed under the methyl p-hydroxybenzoate condition, mounting was also observed under the saline condition. In Experiment 3, preference tests for the odors, as well as for estrous vaginal secretions, were performed. Again, while relative to blanks more attention was paid to the saline and saline + methyl p-hydroxybenzoate swabs that had been inserted into the vagina, no differences in attraction between these two stimuli was noted. Vaginal secretions from estrous donors, however, elicited 2.6 times the interest of the anestrous vaginal inserted swabs and 6 times the interest of the saline or saline plus methyl p-hydroxybenzoate swabs, as would be expected in sexually experienced dogs. These authors concluded that "methyl p-hydroxybenzoate is no more attractive to males than saline alone, when these compounds are applied to the vaginas of anestrous females or presented in our standard odor-testing device" (p. 1503).

It should be reiterated that most if not all of the dogs employed in the 1970 study published in *Science* and all of the dogs employed in Kruse and Howard's study were *sexually experienced*. Hence, even if consistent effects had been observed, they could have been attributed to learning, which is known to markedly influence responses of dogs to estrous urine and vaginal secretions (Doty and Dunbar, 1974). As with the case of copulin, no subsequent studies on this topic have since appeared.

The Estrus-signaling Pheromone of the Rat

Most male rats, like male dogs, spend more time smelling the urine and vaginal secretions of estrous female conspecifics than those of diestrous female conspecifics. While such attraction has been attributed to pheromones located in the secretions of the conspecific female, no pheromone has been chemically identified. As with dogs, this preference is not invariant among male rats and largely depends upon some sexual or social experience on the part of the male in addition to an adequate titer of testosterone (Huck and Banks, 1984; Lydell and Doty, 1972; Hayashi and Kimura, 1974; Taylor and Dewsbury, 1990). Thus, while noncastrated sexually experienced male rats prefer estrous to diestrous urine odor, this is not the case for sexually inexperienced intact males or for either sexually experienced or inexperienced castrated males (Carr et al., 1965; Nakagawa et al., 1981; Stern, 1970; Lydell and Doty, 1972).[3] Mounting without vaginal penetration is sufficient sexual experience to produce the preference (Stern, 1970). Castration has no apparent influence on the ability of sexually inexperienced males to discriminate between estrous and diestrous female rat odors (Carr and Caul, 1962), nor upon their absolute detection threshold for estrous female urine odor (Carr et al., 1962).

The possibility exists that adult male rat preferences for estrous over diestrous female odor may have, in fact, earlier experiential underpinnings than those related to postpubertal sexual encounters. Thus, in a laboratory test situation, a pre-nipple attachment behavior, termed *probing*, is preferentially directed toward anesthetized estrous females relative to anesthetized diestrous females or males during the first few days of life. This occurs even in pups with no nursing experience that are delivered by caesarean section (Fillion and Blass, 1986a) and conceivably reflects similarities between estrous and placental odors. This preference, however, can be overridden by the age of 10 days by applying citral to the nipples and vaginal area of the mother. Thus, pups reared in this scented environment probe only citral-odorized estrous females, not normal estrous females, at that time (Fillion and Blass, 1986a).

To reiterate, no specific chemical or set of chemicals has been isolated from the urine or vaginal secretions of estrous rats that would explain the aforementioned preferences; that is, technically no pheromone has been isolated. This fact, along with the aforementioned dependence upon social and sexual experience, suggests that using the term *pheromone* to describe the preference for estrous over diestrous odor is ill conceived.

The Male Mouse Ultrasonic Courtship Pheromone

Male laboratory mice emit pure-tone ultrasounds of approximately 70 kHz when investigating conspecific females (Sales, 1972; Whitney, 1973). These vocalizations occur in 50- to 300-ms pulses (Sales, 1972; Sales and Pye, 1974) that are nearly continuous during initial courtship and gradually decline over the course of subsequent mating activity (i.e., mounting, intromissions), ceasing after ejaculation (Sales, 1972).[4] Such ultrasonic vocalizations are dependent upon the vomeronasal, not the main, olfactory system for their elicitation (Bean, 1982). Castration decreases the number of vocalizations, whereas testosterone repletion restores them (Dizinno and Whitney, 1977). Dominant males typically emit more ultrasounds in response to females than non-dominant males (Nyby et al., 1976). Although the function of such sounds is not entirely clear, they may serve to signal sexual, rather than agonistic, intentions of the male, mimicking ultrasounds emitted by infant mice that attract females and possibly reduce their aggression (Whitney et al., 1973; Nyby, 2001). In rats, analogous vocalizations have been associated with not only appetitive aspects of sexual behavior (Barfield et al., 1979), but with juvenile play (Knutson et al., 1998), high arousal (Knutson et al., 1999), and anticipation of rewards (Burgdorf et al., 2000), conceivably reflecting a positive affective state (Knutson et al., 1999).

The source of this "ultrasound releasing pheromone" has proven illusive. Soiled bedding from female-occupied cages can elicit such vocalizations in some males, as can cotton balls soaked in female, but not male, urine (Whitney et al., 1974; Nyby et al., 1977). Anesthetized females wrapped in odor-impermeable plastic bags elicit ultrasounds when either the fronts or the backs of their bodies are exposed to the males, as do cotton swabs that have been rubbed on their faces and cheeks (Nyby et al., 1977). Vaginal secretions, as well as female saliva, can similarly elicit such behaviors (Nyby et al., 1977; Byatt and Nyby, 1986). However, cotton swabs wiped on the faces and cheeks of males produce no more ultrasounds than clean cotton swabs (Nyby et al., 1977).

In the initial studies of this phenomenon, the role of sexual or social experience in the elicitation of the ultrasounds was ambiguous. Thus, Nyby et al. (1977) reported that while such experience appears to be critical for DBA/2J male mice, this may not be the case for C57BL/6J × AKR/J hybrid mice. They found that the latter mice, without "heterosexual experience beyond 21 days of age," still exhibited ultrasound vocalization to female stimuli, suggesting that "different genotypes require different amounts of social experience to discriminate between 'maleness' and 'femaleness'. Potentially even more interesting from an evolutionary point of view is the possibility that inbreeding depression leads to a need for greater amounts of social experience to recognize sexual cues" (p. 341).

All of the hybrid males used in the latter study, however, had been tested once with an adult female 48 hours prior to the urine tests, and subsequent studies by this group found that a single brief social encounter with a female is all that is necessary for subsequent ultrasound elicitation. Thus, Dizinno and Whitney (1978) demonstrated that both DBA/2J in-bred and AKD2F1/J hybrid male mice required social interaction, in adulthood, with adult females (and perhaps adult males) before female mouse urine elicited ultrasounds. In this study, social interaction con-sisted of exposure of the mice to a female A/J mouse for 3 minutes, and exposure to male A/J mouse for 3 minutes, in orders counterbalanced among subjects. In no case did "any subject engage in a full mount or in copulation with a stimulus animal." The authors concluded that:

> The results of this experiment demonstrate that both the acquisition and the maintenance of the male mouse ultrasonic response to female mouse urine are influenced by experiential factors. First, it appears that an adult male requires some experience with adult conspecifics before female urine acquires its ultrasound-eliciting properties. Second, repeated presenta-tions of female urine to a socially experienced male, in the absence of the female, causes the urine to cease eliciting ultrasounds. Sexually naïve males as well as socially experienced males that have "extinguished" their re-sponse to female urine do emit ultrasounds upon first subsequent expo-sure to a female. These results are consistent with the hypothesis that the salience of female urine is learned during adulthood, and that the acquisi-tion of this "pheromonally mediated" behavior may be an instance of classical conditioning. (p. 108)

That same year, Nyby et al. (1978) demonstrated that perfume elic-ited high levels of ultrasonic vocalizations in male mice (C57BL/6J × AKR/J hybrids) when such males had a prior brief encounter (less than 3 minutes without mounting) with an adult perfumed female. Males that had been exposed to the perfume only as weanlings in the litter situation exhibited no such responses. The presence of the perfume in the litter situation, as well as on the female briefly encountered as an adult, pro-duced the strongest ultrasound elicitation. Urine elicited ultrasounds in males that had briefly interacted with females as adults, regardless of whether or not the females had been perfumed. However, urine did not elicit ultrasonic vocalizations if the males had no adult social encounter with a female. A similar set of findings was noted in a study in which the females encountered in adulthood were odorized with female rat urine or with urine from female mice that had consumed fenugreek, a spice known to alter urine odor (Kerchner et al., 1986). Nyby et al. (1978) concluded, "We have demonstrated (a) lack of specificity in the female odors that elicit ultrasounds and (b) that the chemical elicitation of adult ultrasounds is heavily influenced by adult experience. While some mammalian chemical communication systems might exist which can be classically (appropriately) described as 'pheromonal,' our results clearly demonstrate a mammalian system that is not consonant with the classical pheromonal concept" (p. 550).

Several studies sought to determine the metabolic pathways responsible for the aforementioned ultrasound-eliciting "pheromone," with the ultimate goal of chemically characterizing it. Since the ultrasonic "calling" to a female's urine odor by a male briefly exposed to the female is eliminated if the female donating the urine is hypophysectomized, the putative pheromone was first thought to depend upon an intact pituitary gland. However, this proved naïve. Thus, if the first exposure of the male was to a hypophysectomized female, the male subsequently emitted ultrasonic vocalizations to urine from a hypophysectomized female (Maggio et al., 1983), implying that salient odor learned at the time of the encounter is the critical factor.

If pheromones are, in fact, involved in this behavior, then the "pheromone" of the hypophysectomized female must be different from the "pheromone" of the intact female, as it is not dependent upon an intact pituitary gland. Furthermore, any factors that alter the odor of the female (e.g., diet) would similarly come into play to produce any number of "pheromones." Importantly, learning need not be confined to odors, as even a plastic bag, or in some cases the entry of a human into a colony room, may acquire the property of eliciting ultrasounds, leading Nyby (2001) to conclude that "almost any stimulus that allows a male to anticipate an encounter with a female would appear to acquire the property of eliciting ultrasounds" (p. 10). These observations argue for an important role of learning of whatever salient stimuli may identify the female at the time, rather than for a specific pheromone.

While the aforementioned studies strongly suggest an important role for experience in the elicitation of ultrasounds, Sipos et al. (1992) subsequently reported that freshly voided urine elicits ultrasounds in male mice that have had no adult heterosexual interactions. This phenomenon is said to depend upon both the main and the accessory olfactory systems. Unlike the earlier studies, in which the urine was collected from metabolism cages, the urine in these studies was collected by grasping a female mouse by the loose skin of the dorsal neck and, if needed, by palpating the bladder. The ultrasound-eliciting activity of urine collected in this manner disappeared after storage for 15 to 18 hours in a sealed syringe, a loss that was preventable by adding antioxidants to the mixture.

These findings were interpreted as reflecting the presence of an "ephemeral sex pheromone" whose degradation is oxygen-dependent and whose elicitation does not require heterosexual social experience on the part of the male. However, such an agent has never been chemically identified and alternative explanations are possible. For example, the fresh urine used in this study was collected from a mouse that most likely was under considerable stress, making it possible that the stimulus for eliciting ultrasonic calling could be an odor associated with stress, that is, adrenocortical activation. As in the case of the hamster, the degree to which earlier social contacts with litter mates influences the induction of this response is not known. Even if one accepts the findings of this study prima facie, one still has to question its generality and importance in establishing the presence of a "pheromone" in light of the aforementioned pre-potent role of learning in most analogous situations.

Phenylacetic Acid, the Major Ventral Scent-marking Pheromone of the Male Mongolian Gerbil

Mongolian gerbils, *Meriones unguiculatus,* have a sebaceous gland on their ventrum used to mark sectors of their living space. In males, marking with this structure has been linked to exploration, social dominance, and territoriality, whereas in females it has been associated with exploration and maternal behaviors (Thiessen et al., 1970; Wallace et al., 1973; Owen and Thiessen, 1974). In a 1974 study published in *Science,* the volatile components of the ventral gland sebum were sampled and, using thin-layer and gas chromatography, phenylacetic acid was found to be the major volatile of the odorous component (to humans) of the secretion (Thiessen et al., 1974). Behavioral bioassays consisted of a conditioned suppression test and a stimulus exploration test.

In the conditioned suppression test, the animals were first trained to avoid an electric shock by suppressing bar pressing in the presence of ventral scent gland sebum. The degree to which suppression generalized to the phenylacetic acid was measured. Clear suppression was observed. In the exploration test, the subject was placed in a test box through which odorized or non-odorized air was flowed. A response was defined when the subject approached the entry point of the airstream, reached up, and sniffed. Responses were averaged across 2 5-minute control periods and 2 5-minute stimulus periods counterbalanced in order within a 20-minute test session.

The mean exploration responses are shown in Table 5.2. Note that the amount of exploration was significantly lower for the phenylacetic acid than for the sebum in two of the three phenylacetic acid series cases (compare top two values with asterisks to bottom three values under stimulus column; t-tests, ps ≤ 0.002). Note also that among the largest observed effects were those between the controls of the sebum group (top three entries in control column) and the controls of the phenylacetic acid group (bottom three entries in control column; all ps < 0.001), and that the responses to phenylacetic acid (bottom three values of stimulus col-

TABLE 5.2. *Exploratory Responses of Gerbils Exhibited during Two 5-minute Control and Two 5-minute Stimulus Exposure Conditions*

Test stimulus	Sample size	Mean (SEM) number of responses		t-value	p-value
		Control	Stimulus		
Sebum series 1 (from 2 males)	10	3.5 (0.75)	9.0 (1.34)*	5.04	< 0.01
Sebum series 2 (from 2 males)	10	2.1 (0.72)	3.6 (0.61)	2.77	< 0.05
Sebum series 3 (from 4 males)	8	2.9 (0.69)	7.8 (0.62)*	4.97	< 0.01
Phenylacetic acid series 1	10	0.8 (0.29)	3.8 (0.61)	6.66	< 0.01
Phenylacetic acid series 2	10	0.4 (0.22)	3.7 (0.54)	4.78	< 0.01
Phenylacetic acid series 3	10	0.5 (0.16)	3.6 (0.36)	9.68	< 0.01

Source: Modified from Thiessen et al. (1974).
Notes: Ventral gland sebum was obtained from cotton swabs saturated with sebum collected from either two males or four males. Controls were "fresh air." Asterisks denote the fact that sebum produced significantly greater responses than phenylacetic acid in two of three sebum series groups. Numbers in parentheses presumably reflect standard errors of the means (SEMs).

umn) did not differ significantly from responses to the sebum series controls (top three values of control column) (all ps > 0.05). If one assumes the validity of the control cases as a baseline within each of the six groups, then, paradoxically, the phenylacetic acid was five to nine times more effective, and the sebum stimulus only two to three times more effective, than the control baseline.

This study is a good example of the difficulties in isolating single components, that is, "pheromones," from a complex mixture and demonstrating equivalent responses to the original secretion. Like many analogous studies, this research has not stimulated subsequent work.

2-(*sec*-butyl)thiazoline, 2,3-dehydro-*exo*-brevicomin, and Major Urinary Proteins as the Aggression-eliciting Pheromones of the Male Mouse

Sexually mature male mice reliably attack strange non-castrate male mice placed into their cages. In general, female and castrate male intruders receive many fewer attacks than gonadally intact males, a phenomenon that is mitigated by injecting the castrates with testosterone (Mugford and Nowell, 1971). Urine and preputial secretions are key components of the involved stimulus complex, as evidenced when these materials are smeared on castrate males or females (Mugford and Nowell, 1971; Heyser et al., 1992; Mugford and Nowell, 1972; Lee and Brake, 1971; Lee and Griffo, 1973). In laboratory settings, a dose-response relation exists between the amount of testosterone injected into a urine donor and the level of attack behavior (Mugford and Nowell, 1972). Deodorized males, as well as males receiving the anti-androgen cyproterone acetate, receive fewer attacks than intact males (Lee and Brake, 1971; Jones and Nowell, 1974a; Nowell and Wouters, 1973). The vomeronasal organ (VNO) is implicated in such behaviors, since genetically modified male mice lacking the TRP2 cation channel, which is specific to vomeronasal receptor neurons, do not initiate such attacks (Stowers et al., 2002). Female mice lacking the TRP2 channel that are lactating also fail to elicit aggression toward a male intruder (Leypold et al., 2002).

The claim has been made that male mice produce specific pheromones, which have been chemically isolated, that elicit the aggressive behavior from other males (Mugford and Nowell, 1970). One group has reported that the involved pheromone consists of two testosterone-dependent urinary volatiles, 2-(sec-butyl)thiazoline and 2,3-dehydro-exo-brevicomin (Novotny et al., 1985). These agents, associated with increased aggressive responses among males in standard behavioral bioassays, do not work when added to water, but appear to be synergistic when added to castrate male urine, implying they require the context of other chemicals or substrates to be effective. Neither of these agents fully mimics the responses seemingly induced, in laboratory conditions, by the male urine or preputial glands.

Another group believes the protein component of the major urinary protein (MUP) complex is the key player in the initiation of aggressive responses (Chamero et al., 2007). The MUP complex accounts for most of the proteins found in both male and female mouse urine, reaching levels as high as 30 mg/ml in the male, a figure 3 to 4 times the level found in female mouse urine. These agents are synthesized in the liver, secreted into the serum, and excreted by the kidneys, being the product of a multigene family of approximately 30 genes and pseudogenes found on chromosome 4 (Clark et al., 1985). However, when male mouse urine was fractionated into low and high molecular weight components, aggression-promoting activity remained in both fractions, suggesting multiple chemicals are involved. Whole urine appears to produce longer duration fighting than either fraction alone, although in the small sample of six mice this effect was not statistically significant ($p = 0.10$) (Chamero et al., 2007).

On the surface, the notion of a simple aggression-promoting pheromone largely dependent upon the VNO seems straightforward. However, aggressive responses in mice are complex and preclude a classical pheromonal or stimulus-based interpretation, being influenced in nature by early experience, crowding, group size and composition, resource availability, and seasonal changes (King, 1973). If a "pheromone" were truly involved in eliciting agonistic behaviors, one would expect relatively invariant and stereotyped responses toward the stimulus. A number of observations show this is not the case. First, mature male mice are not continuously displaying aggressive behaviors toward their own secretions, as would be expected if context, novelty, and other factors played no role in the mediation of the response. Second, adult male mouse urine placed on mouse pups does not elicit aggressive responses from other male mice; in fact, such urine actually reduces infanticide (Mucignat-Caretta et al., 2004). This reiterates the importance of context and demonstrates that aggressive responses are not invariably tied to the stimulus. Third, aggression is minimal among group members within demes, even though a number of the members of the group—most significantly the dominant male or males—have high testosterone titers and would be expected to be producing the aggression-eliciting pheromone. This lack of aggression depends, in part, on the development of a social dominance hierarchy and the scent-marking behaviors of the dominant males that familiarize the other deme members with their odors. Fourth, exposing a male mouse for 1 hour daily for 10 days to soiled wood shavings from the cage of a strange male mouse eliminates, for the most part, agonistic behaviors directed toward that male in subsequent encounters. Such exposure, which unlikely reflects sensory adaptation given the physiology of the vomeronasal system (Luo et al., 2003), also mitigates, to a lesser extent, agonistic behaviors directed toward other mature male conspecifics (implying that familiarity with any conspecific male odor in this context decreases agonistic tendencies) (Kimelman and Lubow, 1974). It would seem that it is the strangeness of the odor, or by inference the individual, which most likely contributes to the elicitation of aggressive behaviors.[5] If agonistic behaviors were being mediated by a pheromone, then one would be inclined to postulate that each mouse has a different aggression-eliciting pheromone. Fifth, the argument that MUPs are the basis of the aggressive responses overlooks the fact that such agents are also produced by females, albeit at lower levels, which are not normally attacked. An alternative explanation is that MUPs alter individual recognition cues so that the stimulus bearer is perceived as strange or unfamiliar. Finally, the existence of aggression-eliciting pheromones makes little sense from an evolutionary perspective. Brain et al. (1987) note the following:

> Evans (1979) claimed that there is no logical way of arguing for the existence of a "pheromone" whose function is to release aggressive behaviour from other mice. This, to all intents and purposes, would involve a signal meaning "attack me please" and would result in a loss of fitness on the part of the signaling individual. The results of the present studies support this position and suggest that the odours really act as personal labels signaling «I am a threat to you». For example, the odour of preputial sebum

of dihydrotesterone-treated castrates probably identifies the donor as «an unfamiliar, mature, sexually active, territorial male» to non-habituated conspecifics who respond accordingly. In this way there is no loss of fitness. (p. 283)

Even if one accepts 2-(*sec*-butyl)thiazoline, 2,3-dehydro-*exo*-brevicomin, and some combination of MUPs as salient elements of a stimulus associated in the elicitation of agonistic activity in some conspecific male mice, can they be viewed as pheromones in light of the aforementioned issues? Importantly, no single agent uniquely elicits such activity. A seemingly more parsimonious explanation would be that these agents are part of a stimulus complex that becomes interpreted as threatening, perhaps as noted above because of strangeness or novelty, a concept with considerable empirical support (Mackintosh and Grant, 1966; Alberts and Galef, 1973). As noted by Alberts and Galef (1973) for rats, familiarity is a key component in the initiation of aggressive responses: "the response of wild Norway rats to conspecifics is determined by a multitude of stimuli perceived via several sensory modalities. Response to a conspecific as such (amicable and sexual behavior) can occur in the absence of olfactory inputs. On the other hand, the initiation of aggression would appear to be dependent on olfactory stimuli arising from an unfamiliar individual. Both the duration and direction of aggressive behavior is further modified by the behavior [e.g., movement] of target animals" (p. 242).

The Maternal Pheromone of the Rat

In a series of innovative studies, infant rats of the Wistar and Sprague-Dawley strains were found to be attracted to a maternal odor, labeled the "maternal pheromone," during the second through the fourth weeks of age (Leon and Moltz, 1971, 1972; Leidahl and Moltz, 1975; see also Holinka and Carlson, 1976; Nyakas and Endröczi, 1970). A similar phenomenon has been reported for house mice (Breen and Leshner, 1977). The main source of the stimulus is the cecotroph portion of the maternal anal excreta.

It is now known that this phenomenon is strain dependent and influenced by learning (Leon, 1975; Coopersmith and Leon, 1984). Such learning is facilitated by tactile stimulation, such as from a mother rat licking her pups (Pedersen and Blass, 1982). Importantly, the attractiveness of the cecotroph has been found to depend upon the mother's diet and cecal bacterial populations, as well as the production of bile and prolactin (Leon, 1974, 1975, 1978; Moltz and Leidahl, 1977). Thus, antibiotics that eradicate the bacterial flora eliminate the attractiveness of the excreta. Pups raised with mothers on a particular diet are attracted to the odor of mothers eating that specific diet (Leon, 1975). Galef (1981) demonstrated, in a now classic experiment, that the attractiveness of a dam's feces varied between animals fed two slightly different formulations of Purina Laboratory Chow No. 5001 which, according to the manufacturer, differed subtly in only a few constituents.

The problems with the maternal pheromone concept are well illustrated by an extensive series of studies performed by Clegg and Williams (1983). These investigators approached their project under the assumption that a pheromone is an entity unique from that of simply an odor. In the first of 8 experiments, the age of the pups (18 days versus 24 days), duration of maternal deprivation (3 hours versus 18 hours), and type of test stimuli (e.g., own mother versus a virgin female; own mother versus an adult male) were assessed. No evidence of a statistically meaningful preference was noted for the pup's own lactating mother over the other stimulus animals. Unlike in earlier studies, these authors eventually employed anesthetized rats as stimulus objects for the following reason:

> This procedure [i.e., use of unanesthetized female stimulus animals] was initially followed in the present study. Eighteen PVG/C hooded rats aged 18 days were given the choice of their own mother as opposed to a virgin female of the same age and strain. Pups underwent 3 hr of pretest maternal deprivation. All of the pups entered the goal compartment containing their own mother. But it was apparent that factors other than a pheromonal agent had helped to bring this about. As soon as a pup was placed in the apparatus the mother became extremely agitated, possibly as a result of ultrasonic calling by the pup (e.g., Allin and Banks, 1972; Smith, 1975) and in moving about emitted auditory cues. This was in sharp contrast to the non-lactating females, which tended to settle down in the goal box and go to sleep. Any maternal attraction mediated by olfactory cues was thus confounded by auditory cues and maternal retrieving. Mobile

live stimuli can therefore not properly be used to assess maternal pheromone phenomena. (p. 225)

In the second study of the series, Clegg and Williams tested excrement collected from lactating versus virgin females. The amount of the fecal material (3 g) was equated across groups, since lactating females typically produce more excrement than non-lactating females, thereby potentially producing a confounding factor of stimulus quantity. No evidence for a meaningful preference for the excrement of the lactating females was observed in either the Wistar or Sprague-Dawley rats, although this was not the case for the PVG/C strain rats. Of 151 PVG/C rats, 79 chose the goal box containing the maternal excrement, 47 chose the side containing the virgin female excrement, and 25 remained in the arena, not moving into either compartment. While the number of PVG/C animals choosing the side of the maternal excrement was significantly higher than the number that chose the side of the virgin female excrement, the large number of pups that chose the virginal excrement, in combination with those that did not make any choice at all, led these authors to the conclusion that the effect, at best, is weak.

The remaining experiments examined such factors as air flow rates in the testing apparatus, maternal deprivation times, olfactory function in the pups, different types of laboratory food, and the effects of presenting cecal contents obtained from sacrificed dams in an effort to observe a robust phenomenon. The data suggested (1) pups learn to identify olfactory cues associated with the diet, and (2) cecal contents produced an approach behavior relative to an empty goal box. In relation to the latter observation, the authors write, "So once more there is a clear suggestion of olfactory cues influencing behavior, but again without the control over that behavior that would be expected on the maternal pheromone hypothesis" (p. 234). Clegg and Williams reiterate the importance of Galef's observations concerning subtle elements of diet and note the following about pheromones in their discussion:

> The implications of Galef's discovery are indeed far reaching, and it highlights the problem of deciding what does or does not constitute a pheromone. The pheromone concept is employed to go beyond the notion that animals use olfactory cues. It carries the implication that the stimulus controls behavior by eliciting a stereotyped and reliable response within

any particular species (Karlson and Lüscher, 1959). If certain strains within a species fail to demonstrate the appropriate response, or if members of the species do so under highly specific conditions, then can the effect be called "pheromonal"? (p. 234)

(Z)-7-Dodecen-1-yl Acetate, the Sex Pheromone of the Female Asian Elephant

During periods when they are not eating, the body region most commonly contacted by the trunk of captive Asian and African elephant herd members is the ano-genital region (Rasmussen and Schulte, 1998). A few weeks prior to ovulation, female elephants appear to deliberately urinate in the presence of male elephants or after exposure to their musth urine (Rasmussen and Krishnamurthy, 2000). The preovulatory female urine is said to contain a pheromone that elicits a high frequency of flehmen in the male, as well as other responses associated with mating (e.g., placing the trunk tip on a urine sample) (Rasmussen et al., 1997). Using the flehmen response as the bioassay and extraction of components from estrous urine, (Z)-7-dodecen-1-yl acetate (Z7–12:Ac) was isolated, purified, and deemed the sex pheromone of the elephant (Rasmussen et al., 1997). This material was clearly shown to be greater in urine sampled from females in the early follicular phase just prior to ovulation. Z7–12: Ac has been reported to be a common sex attractant in "pheromone blends" of more than 126 species of insects, particularly those within the Lepidoptera (Rasmussen et al., 1996).

While this agent, in both its extracted and synthetic forms, increased flehmen responses in male Asian elephants living in isolation from their peers in an American zoo, the magnitude of the responses was typically lower than that to whole estrous urine and monotonic dose-response relations were not present. The mean flehmen responses per hour observed in 7 trials for each of 9 male elephants for the most active concentrations of Z7–12:Ac extracted from urine was 2.05. In contrast, the mean response to whole preovulatory urine was 7.00 per hour for all trials. In 4 male elephants exposed to synthetic Z7–12:Ac, the mean responses per hour for the first 3 trials was 1.99 and the subsequent 7 trials was 1.87. Nevertheless, in trials with 1.0 mM of the pure Z isomer using 5 additional males, the response rate was considerably higher (5 flehmen responses per hour). Interestingly, a synthetic novel agent used as a control, 2-n-propyphenol (10 mM), elicited 4 flehmen responses per hour in this study, in accord with observations that novel odors can elicit increased flehmen responses. Flehmen responses to novel odors, however, reportedly habituate more rapidly than responses to estrous urine.

Rasmussen et al. (1997) performed additional studies with Asian elephants that worked in lumber camps of Burma (Myanmar). In contrast to the aforementioned solitary living animals, the dominance position and general sexual prowess of these elephants were known from both written records and reports of the Burmese rider-handlers who managed and cared for the elephants during much of their lives. The number of flehmen responses to synthetic $Z7–12$:Ac was much higher than that observed in the American captive elephants. Thus, the Myanmar elephant flehmen responses ranged from 2.33 to 10.00 responses per hour, compared to the 0.43 to 3.00 responses per hour of their American elephant counterparts. No flehmen data based on whole urine presentations to the Myanmar elephants were presented. However, experience is likely critical in determining the number of flehmen responses directed toward the $Z7–12$:Ac, as indicated by the following summary of findings regarding these working elephants:

> Fascinatingly, the examination of the social status of the males revealed that all of the responder males were dominant animals who either had ready access to females or were known to be proven breeders. Conversely, all the animals that either backed away from the sample or ignored it were the younger subdominant males. Clearly, the context and the societal structure of the elephant needs to be considered carefully in further studies of $Z7–12$:Ac as a sex pheromone. (p. 434)

Frontalin (1,5-dimethyl-6,8-dioxabicyclo [3,2,1] octane), the "Multipurpose" Male Elephant Pheromone

Adult male elephants produce increased levels of secretions from their temporal glands when in musth, a heightened state of breeding readiness when testosterone levels are high. Analysis of these secretions has revealed the presence of ketones and alcohols, as well as a bicyclic ketal compound frontalin (1,5-dimethyl-6,8-dioxabicyclo[3,2,1] octane). This substance is also present in the urine and breath of mature males in musth (Rasmussen, 1998), but is absent from the temporal gland secretions of subadult male elephants during their first "moda" musth (Rasmussen et al., 2002). Frontalin is said to function "as a male-generated pheromone in the Asian elephant, eliciting a range of behaviors in conspecifics depending upon their sex, age and physiological state" (Rasmussen and Greenwood, 2003). On the other hand, "Tests with subadult and adult male and female African elephants evoke only sniffs and exhalations and no overt behaviors" toward frontalin (p. 442).

In the defining behavioral study (Rasmussen and Greenwood, 2003), frontalin was the focus of interest largely because it is "a well-characterized bark beetle pheromone" and "the ketonic compounds elicited variable behavioral responses" (p. 434). Six adult males, three subadult males, four pregnant females, six adult females in the follicular phase of their estrus cycle, and six adult females in the luteal phase of their estrous cycles served as subjects. Two stimuli, 100 μM of synthetic frontalin in a phosphate buffer and buffer alone, were randomly placed a minimum of 20 feet apart on the concrete in a test area. The concentration of the frontalin was similar to that contained in the temporal glands.

The males were tested individually whereas the females were tested in groups. The elephants' responses to the frontalin were categorized into "main olfactory," "pre-flehmen," and "flehmen" categories. The main olfactory responses were those of sniffs, trunk shakes, and forceful exhalations or blows. The pre-flehmen responses consisted of checks (where the trunk tip finger is placed in the sample area), places (where the entire trunk tip is flush with the sample area), and sucks (where the place response is associated with sucking). The flehmen response occurs following the elephant's placement of the material by the trunk into the openings of the vomeronasal duct on the anterior dorsal palate. Since the elephants rarely exhibited a sniff or check to the control stimulus, only responses to the frontalin were presented in Rasmussen and Greenwood's paper.

The primary results are summarized in Table 5.3. It is apparent that

TABLE 5.3. *Median (Interquartile Range) Responses per Hour of Elephants to Synthetic Frontalin on First Presentation Trial*

Subject group	Main olfaction (sniffs, trunk shakes, blows)	Pre-flehmen (checks, places, sucks)	Flehmen
Adult males (n = 6)	1.00 (1.00–2.00)	1.50 (0.00–2.00)	0.00 (0.00–1.00)
Subadult males (n = 3)	4.00 (3.25–4.00)	9.00 (7.50–11.25)	2.00 (2.00–2.75)
Follicular phase females (n = 6)	3.50 (3.00–4.00)	5.60 (5.00–6.00)	0.50 (0.00–1.00)
Luteal phase females (n = 6)	1.50 (0.00–2.00)	0.00 (0.00–0.00)	0.00 (0.00–0.00)
Pregnant females (n = 4)	1.00 (1.00–1.50)	7.00 (4.50–10.0)	0.00 (0.00–0.00)

Source: Data from Rasmussen and Greenwood (2003).

the frequency of responses directed toward frontalin was relatively low, with pre-flehmen responses predominating. Most of the pre-flehmen responses were exhibited by subadult males and by pregnant and follicular phase females. Overall, the subadult males seemed to exhibit more responses of all types than the other groups to this agent.

Based on these data, the authors view frontalin as carrying "an intra-specific message to conspecifics, eliciting immediate chemosensory and behavioral responses" and fulfilling Meredith's (Meredith, 2001) phero-mone criterion of providing "evolutionary fitness" and "mutual benefit" for both sender and receiver." The investigators state that

frontalin as a pheromone also imparts important short-term social and reproductive messages that elicit immediate and evidential reactions and behaviors from both sexes. Older males exhibit awareness of frontalin by sniffing or checking, but only indifference follows. In contrast, younger subadult males proffer the same frequency of sniffs to frontalin but significantly more pre-flehmen and flehmen responses, and often vocalize or exhibit a variety of retreat postures. Females also exhibit varied immediate responses that segregate according to their individual hormonal condition and reproductive state. Luteal-phase females, while not disinterested in frontalin, have a much lower response than follicular-phase females in all categories of chemosensory responses. While the olfactory responses of luteal-phase and pregnant females were similar in frequency, pregnant females were very responsive in the pre-flehmen category exhibiting many check, place and suck responses, enabling identification of frontalin and probably its concentration. (pp. 442–443)

These authors stress that learning and other factors are involved in producing the observed responses:

> In examining pheromonal actions in elephants, their high cognitive abilities and excellent long-term memories, especially regarding odors, need to be considered. For a meaningful study of captive male elephants, a personal life experience history, including past and present musth status, hormonal levels, and past associations that account for individual differences, is necessary to discern possible functions. For example, one [adult] male, although of similar age to the other older adult males, is subdominant and is less socially mature as the result of his particular life experiences; consequently, his response for the first two [frontalin] assays fit the pattern of a younger male. As the dominance dynamics of the males changed, his responses fit into the adult grouping. (p. 444)

More recently this group presented behavioral data from actual male temporal gland secretions in a paper whose main point was that the median ratio of (+) and (−) enantiomers changed from 60:40 in younger males to 50:50 in older ones (Greenwood et al., 2005). The behavioral responses of apparently the same subjects represented in Table 5.3 were presented as means, not medians, and categorized into attraction and avoidance responses. In this new categorization scheme, *attraction responses* included sniffs, trunk shakes, places, and flehmens, whereas *avoidance responses* included sample circling, ears erect, backing up, audible forceful exhalation, foot scuffling, running away, and accompanying vocalizations of trumpeting and roaring. No subjects except for follicular phase females evidenced any attraction responses to the secretions, which is not only counterintuitive (since one would assume some sniffing to the whole secretion would be a prerequisite for avoidance) but differs qualitatively and quantitatively from the responses noted in Table 5.3 to frontalin, where sniffs, trunk shakes, blows, checks, places, and sucks were observed in all groups, most notably the young males. The mean "attraction" and "avoidance" responses estimated from Figure 1d of their paper are shown in Table 5.4.

Given that (1) infrequent responses are directed toward frontalin per se, (2) responses to the whole secretion appear to differ from those to frontalin alone, (3) many other volatile and nonvolatile chemicals are present in the subauricular gland, and (4) learning and other factors play

TABLE 5.4. *Estimated Mean Values of "Attraction" and "Avoidance" Responses*

Subject Group	Mean (SEM) attraction responses (sniffs, trunk shakes, places, flehmens)	Mean (SEM) avoidance responses (circling, ears erect, backing up, exhalation, foot scuffling, running away, trumpeting/ roaring)
Adult males (n = 6)	None	1.12
Subadult males (n = 3)	None	0.50
Follicular phase females (n = 6)	3.90	None
Luteal phase females (n = 6)	None	0.76
Pregnant females (n = 4)	None	0.44

Source: Responses derived from Greenwood et al. (2005), Figure 1d.

an important role in determining the responses to frontalin and the parent secretions, should this single chemical be considered a "multipurpose" pheromone in the Asian elephant?

The Pheromone of the Subauricular Gland of the Male Pronghorn Antelope

Like the territorial male Thomson's gazelle shown in Figure 4.3, the male pronghorn antelope, *Antilocapra americana*, employs subauricular glands located below each ear to mark areas of vegetation. Using short-path distillation, gas chromatography, mass spectrometry, and other techniques, eight major components of the glands were identified in 1974 (Müller-Schwarze et al., 1974). These were (1) 2-methylbutryic acid, (2) isovaleric acid, (3) 13-methyl-l-tetradecanol, (4) 12-methyl-l-tedradecanol, (5) 13-methyltetradecyl 3-methylbutyrate, (6) 12-methyltetradecyl 2-methylbutrate, (7) 13-methyltetradecyl 2-methylbutyrate, and (8) 12-methyltetradecyl 2-methylbutyrate.

Behavioral responses to these compounds were assessed in three hand-raised adult captive male pronghorns (Müller-Schwarze et al., 1974). These animals were repeatedly tested over the course of eight months for the intensity of their sniffing, licking, thrashing, and marking responses to four different concentrations of the aforementioned stimuli (0.1, 1.0, 10, and 100 μg), as well as to blanks and to the whole secretion. Responses to mixtures were also assessed by adding individual compounds in order of increasing and decreasing boiling points with the relative proportions of acids, alcohols, and esters corresponding to those found in the "scent on the hair overlying the subauricular gland." The tests were performed in 15-minute test sessions in which 6 stimuli were presented at a time, one of which was a blank and the other petroleum ether. The stimuli were coated onto 15 cm × 1 cm horizontal rods positioned atop 80 cm tall nylon sticks located 50 cm apart within a 24 square m triangular pen. Responses were also recorded to stimuli emanating from permanent wooden posts that were commonly marked with subauricular scent but were never exchanged or cleaned.

The mixture containing all of the eight compounds fell short of inducing the same intensity of responses than that of the whole subauricular gland. Isovaleric acid alone produced more intense responses than the other stimuli at all four concentrations tested, including the whole secretion. The intensity of the responses given to the mixtures was also enhanced with the addition of this compound. Five of the seven remaining compounds also elicited, at some concentration, more intense responses than those elicited by the blanks, but never at the level observed for isovaleric acid. Interestingly, the magnitude of responses to the whole gland secretion was greater when the stimuli were presented on the permanent vertical wooden posts than on the nylon rods, suggesting that they may have been influenced by such factors as prior odors on the posts (since they were never cleaned), visual context, familiarity of the sign post, the height of the stimuli relative to the ground, or differential diffusion of the stimuli from the different sources.

Unfortunately, this study suffers from some basic shortcomings and has never been replicated. First, the response measure was a complex weighted average of the sniffing, licking, thrashing, and marking responses. As indicated by the authors, "The exact number of points for each behavior was determined by the response to blanks that were presented with the samples of a particular test series. For example, in one particular series, in response to a total of 24 blanks, the animals sniffed 52 times, licked 27 times, marked 14 times, and thrashed 4 times; points given, therefore, were 1 for sniffing, 2 for licking, 4 for marking, and 13 for threshing. In this series when isovaleric acid was presented at 0.1 µg on 6 rods, the animals sniffed 16 times, licked 9 times, marked 5 times, and thrashed 4 times. The total number of points, thus, was $(1 \times 16) + (2 \times 9) + (4 \times 5) + (13 \times 4)$ for a total of 106" (p. 862). It is clear that this composite weighting can be significantly skewed by a few thrashes. Second, the source(s) of the stimulus materials were not described. Presumably the stimuli came from a number of different males over time. If so, it would be of interest to know their ages and familiarity to the subjects. Third, the entire study is based upon the responses of only three hand-reared animals that were repeatedly tested in many test trials at variable test periods, leading one to wonder about the generalizability of its findings.

As noted by the investigators in the introduction to their paper, the frequency of marking with this gland is influenced by the strangeness of another mark, the novelty of the environment, absence of a dominant male in a bachelor herd, and the return of a dominant male into the bachelor herd. These authors also stated that "the variations usually encountered in field experiments are in part due to known causes, such as diurnal, seasonal, and meteorological factors, even though the pronghorn were tested at the same time of day" (p. 860). Such variation and dependence upon social factors begs the question that, even had the essential chemical elements of the subauricular gland been identified, is it reasonable to consider them "pheromones"?

The Rabbit Nipple-Search Pheromone, 2-methylbut-2-enal

Rabbit kits are dependent upon an olfactory cue on the doe's belly that elicits and guides a stereotyped search behavior for locating the nipple (Hudson and Distel, 1983). Such guidance is important, as a kit must find and attach to a nipple quickly to survive since, in natural settings, the mother is available for nursing only about 3 minutes each day (Coureaud et al., 2000). While at the time of the first suckling episode the nipple may be moistened with amniotic fluid, saliva is apparently not attractive (Hudson and Distel, 1983) and self-grooming does not seem to transfer an attractive agent to body areas distal to the nipple (Coureaud et al., 2001). The source of the odor is unknown, although it appears to be distributed over the nipple epidermis and possibly released from the nipple itself (Coureaud et al., 2001).[6] During the nursing episode rabbit kits switch nipples periodically (average = 2.6 times per minute), a behavior not dependent upon the amount of milk available from any one nipple (Hudson and Distel, 1983). Olfactory bulbectomy eliminates the nipple search behavior and, hence, suckling (Distel and Hudson, 1985). In non-breeding females, the emission of the active stimulus is influenced by day length, peaking in the early summer, and is depressed by ovariectomy and restored by estrogen administration (Distel and Hudson, 1984). Both progesterone and prolactin, probably in concert, increase its emission in estrogen-primed does (Gonzalez-Mariscal et al., 1994), although oxytocin may also be involved (Fuchs et al., 1984). The nipple search response is not dependent upon an intact vomeronasal organ (Hudson and Distel, 1986).

Despite the fact that rabbit kits delivered by caesarean section exhibit normal search and suckling behavior when placed by a lactating doe, Hudson (1999) points out that "this does not exclude the possibility that the response is dependent on prenatal experience of chemical characteristics of the uterine environment. In fact, this might even be considered likely given the steep rise in pheromone emission in late pregnancy (Distel and Hudson, 1984) and reports that rabbit pups are able to learn prenatally odor cues associated with their mother's diet (Bilko et al., 1994; Hudson and Altbäcker, 1982; Semke et al., 1995; Coureaud et al., 1997)" (p. 301). It is noteworthy that when the rabbit mother is perfumed before nursing, the kits learn to respond to the novel odor with the characteristic nipple search behavior in a single brief nursing episode (Kindermann et al., 1991; Kindermann et al., 1994). Interestingly, the opportunity to suck, independent of milk ingestion, is a highly reinforcing element of such learning (Hudson et al., 2002).

Schaal et al. (2003) have reported that the active agent involved in initiating and guiding the nipple search behavior is 2-methylbut-2-enal. Split stream gas chromatography was used to establish behaviorally active volatile peaks from rabbit (*Oryctolagus cuniculus*) milk that can attract and induce oral grasping (Figure 5.1). 2-methylbut-2-enal was found to be the most active agent of those screened and, in a series of studies, met criteria listed by Beauchamp et al. (1976) as to what were most commonly assumed to be essential elements of a mammalian pheromone in the mid-1970s (see Chapter 2, p. 20). The effectiveness of 2-methylbut-2-enal was concentration dependent, with optimal elicitation of the behav-

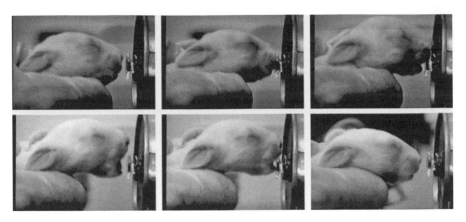

FIGURE 5.1.   Sequence (duration 5 seconds) of a two-day-old rabbit kit's searching–grasping response directed to the glass funnel of a gas chromatograph sniff-port. From Schaal et al. (2003); reprinted by permission from Macmillan Publishers Ltd., copyright 2003.

ior occurring at concentrations 10^{-9} to 10^{-5} g/ml (Coureaud et al., 2004). This material did not elicit the searching and nipple-grasping responses in newborns from another Lagomorph species, *Lepus europaeus,* or from several rodent species (*Rattus rattus, Mus musculus)* or cats *(Felis cattus),* implying species specificity. Freshly obtained colostrum and/or milk from rats, sheep, cattle, horses, and humans did not elicit nipple searching and grasping responses in the rabbit kits and bovine milk was shown not to contain 2-methylbut-2-enal. Kits delivered by caesarean section or that were immediately isolated from their mother after birth clearly responded to the compound, suggesting to the authors that the responses were not learned. Lack of influences of intrauterine exposure to the agent was inferred from the observation that neither blood nor amniotic fluid, when placed on glass rods, produced the behavioral responses in one- to three-day-old pups.

Despite the uniqueness and eloquence of these studies, there is some question whether 2-methylbut-2-enal is the active agent associated with rabbit nipple search behavior. Hudson et al. (2007) point out that (1) rabbit kits successfully search for nipples on sexually receptive and pregnant females long before the start of lactation (i.e., the search behavior is independent of the presence of milk), (2) 2-methylbut-2-enal has not been found on the ventrum of lactating rabbits, (3) concentrations of 2-methylbut-2-enal effective in eliciting nipple search-like responses in rabbit kits are well above physiological concentrations in milk, and (4) several other candidate substances identified by Schaal et al. (2003) in rabbit milk were not tested in the behavioral bioassay.

Late-term or postnatal lack of interest in the smell of blood or amniotic fluid does not necessarily prove that a preference for 2-methylbut-2-enal is *not* learned in some manner *in utero*. Complexity is added by the fact that rabbit kits display strong attraction toward abdominal odors of adult male rabbits, non-lactating female rabbits, and non-lactating non-pregnant female rabbits, although the odor of lactating females is more preferred. The kits' attraction to 2-methylbut-2-enal is not invariant, being influenced by age and degree of hunger. For example, while nearly 100% of newborn rabbit kits deprived of suckling for approximately 5 minutes exhibited searching and grasping responses directed toward a rod containing 2-methylbut-2-enal, only 75 to 80% of those allowed to nurse for approximately 15 minutes prior to such testing did so (Montigny et al., 2006). This effect is more marked in 5- and 10-day-old kits, resulting in post-suckling decreases to nearly 40% of 5-day-old kits and 23% of 10-day-old kits. The pre-suckling responsiveness of these two groups of kits was 87% and 68%, respectively, implying an age-related decrease in responsiveness even under conditions of hunger. Thus, both age and physiological state are critically important in determining the responsiveness toward 2-methylbut-2-enal.

The Erection-eliciting Pheromone of the Rat

Some mature male rats exhibit penile erections in the presence of inaccessible estrous females (Sachs et al., 1994). This phenomenon does not require adult sexual experience on the part of the male and can be induced in the absence of visual and auditory cues by airborne volatiles from the females (Sachs, 1997). The involved airborne stimuli, which have not been chemically identified, appear to be effervescent or very context dependent as bedding soiled by estrous females does not produce this effect (Sachs, 1997). In one study, the odors of fresh urine, but not feces, from estrous females elicited the effect (Kondo et al., 1999), whereas in another the odor of feces from estrous females did so (Rampin et al.,

2006). Although lesions within the medial amygdala, a structure that serves as a major relay of the accessory olfactory system, eliminates the effect (Kondo et al., 1998), removal of the vomeronasal organ does not, in contrast to removing the olfactory bulbs or damaging the olfactory epithelium with zinc sulfate (Kondo et al., 1999). The investigator who discovered this phenomenon stated (Sachs, 1997), "Receptive female rats apparently broadcast a volatile pheromone that promotes erection. Pheromones are well known to attract mates and to act in concert with other stimuli to promote mating. However, this is the first mammalian evidence for a volatile pheromone acting alone to evoke a sexual fixed-action pattern and, in that sense, acting as an airborne aphrodisiac" (p. 921).

One can question whether this phenomenon is best described as being mediated by a pheromone on several counts. It does not occur in all strains of rats and even in those strains where it does occur it is present in only about half of the subjects (Sachs, 1996). For example, in one study of sexually inexperienced Long Evans hooded male rats, only 11 of 20 rats displayed the phenomenon. When another 20 rats were tested and half were given sexual experience prior to testing, essentially the same number (10/20) exhibited the erections (Sachs, 1997). One study found that fecal stimuli collected from estrous foxes and horses, but not males, also induced the effect, contrary to the notion of species specificity that is usually applied in pheromone definitions (Rampin et al., 2006). It is not clear in the latter study as to how many of the 12 subjects actually exhibited the phenomenon. It would be of interest to determine whether lactating female rats that exude a number of hormone-influenced odors also induce penile erection. If so, this response to estrus-related odors may be largely learned early in life.

Male Mouse Sex Pheromones

Nonvolatile compounds in male mouse urine, sensed by the VNO, are attractive to female house mice separated from adult males and their odors from the time of birth. In the defining study on this topic (Moncho-Bogani et al., 2002), pregnant females were housed in a clean room without adult males. On the 19th postpartum day, the pups were sexed and male pups removed. The female pups were subsequently housed in groups of five in a room where no males or their excrement were present. At nine weeks of age, the female mice were tested for their preferences between bedding soiled from multiple males versus clean bedding. Under some conditions, direct access to the bedding was possible (VNO + main olfactory stimulation), whereas under other conditions only bedding vapors could be sampled (main olfactory stimulation only). Initial preferences for the male-soiled bedding occurred only when bedding contact took place, implicating the VNO in the seemingly unlearned preference that is uninfluenced by hormones (Moncho-Bogani et al., 2004). However, when the female mice had prior contact with the bedding, main olfactory system–mediated preferences quickly developed. These and related findings suggested to the investigators that unidentified pheromones exist in the urine that have unlearned rewarding properties that sustain normal sexual behavior and condition main olfactory odor system preferences—conditioning that likely depends upon the basolateral amygdala (Moncho-Bogani et al., 2005).

The VNO-mediation of female attraction to male urine has been confirmed in wild house mice (Ramm et al., 2008) that have greater variation than laboratory mice in the polymorphic MUPs that provide urinary cues implicated in individual recognition (Hurst et al., 2001; Cheetham et al., 2007), inbreeding avoidance (Sherborne et al., 2007), and genetic heterozygosity (Thom et al., 2008). Unlike the defining work, the urine to which the mice were exposed came from individual males, not from a conglomerate mixture of multiple males. Under these conditions, the conditioned preference mediated by the main olfactory system was specific to the urine of the male donor and did not generalize to urine from other male donors. In other words, the learning was not to sex per se, but to the specific male individual that donated the urine (Ramm et al., 2008).

These studies suggest that the VNO may mediate a non-learned preference of female mice for complex sets of chemicals found in conspecific male urine. Whether the attractive agents are dependent upon commensural bacteria, diet, or the richness or novelty of testosterone-based chemicals that are lacking in female or castrate urine is unknown. The VNO-olfactory conditioning appears to be very specific. If a pheromone is involved in this process, then each individual male mouse would seem to need to possess a different pheromone and a search for a specific agent would seem fruitless. It would seem more likely than not that the male-related individuality is encoded by multiple and presumably testosterone-related combinations of chemicals. It is questionable whether such yet-to-be-identified agents, which are likely multiple in nature, highly individualistic, and probably interactive with one another, should be considered pheromones.

Mammalian Alarm Pheromones

A number of animals release odorous secretions when under stress or attack, often as a result of autonomic nervous system reflexes. For example, shrews from the genus *Suncu* become more odorous when being handled (Dryden and Conaway, 1998), humans increase the production of eccrine and apocrine sweat when stressed (Doty et al., 1978), and skunks and pole cats extrude anal sac secretions when frightened (Müller-Schwarze et al., 1998; Lent, 1966; Kurt, 1967). In a number of mammals, piloerection related to defensive reactions increases the release of odorants into the air, since erect hairs facilitate air circulation to the follicles and related sebaceous gland secretions (Flood, 1985). Not surprisingly, many mammals react to stress- or alarm-related odors, exhibiting, for example, fear responses, such as avoidance or freezing (Doty, 1980), and autonomic responses, such as increased body temperature (Kikusui et al., 2001).

In general, mice and rats alter their activity in the presence of odors from stressed conspecifics (Mackay-Sim and Laing, 1980; Dardes et al., 2000; Ropartz, 1966; Zalaquett and Thiessen, 1991; Valenta and Rigby, 1968; Müller-Velten, 2003). Although some investigators describe such changes as being due to pheromones (e.g., Kikusui et al., 2001; Kiyokawa et al., 2004b, 2004a, 2005; Abel, 1991), the responses are highly variable and reflect the specifics of the test environment and previous experience of both the odorant donor and the recipient (King et al., 1975; Thomas et al., 1977). Mackay-Sim and Laing (1981), for example, found that odors from mildly stressed rats (corticosterone levels ~50 μg/100 ml plasma) produced more locomotor activity in male conspecifics than odors from non-stressed rats (corticosterone levels ~20 μg/100 ml). In contrast, odors from highly stressed rats (corticosterone levels ~95 μg/100 ml) produced less activity than did odors from non-stressed rats. No evidence of avoidance or attraction to the stressed odors was observed, although in an earlier study using a different test paradigm odors from mildly stressed rats were found to be somewhat attractive to males (Mackay-Sim and Laing, 1980). Using foot-shocked BALB/cJ mice as odor donors, Zalaquett and Thiessen (1991) noted that when recipient mice were in a tube, they tended to avoid the end of the tube in which such odors were introduced. However, when in their home cage with other mice present, they often sought out the source of the odor, exhibiting increased activity, rearing, and sniffing of the air—behaviors that habituated over time. They concluded that "a restriction of behavioral opportunities will lead to odor avoidance; however, when the environ permits, the behavioral reaction to odors becomes more complex" (p. 221).

It is of interest that conditioning can influence the nature of the odors elicited by a stressed conspecific. For example, running speeds in a runway are decreased in rats that follow an odor trail of a previous rat "frustrated" by the lack of a food reward on a previous trial (Ludvigson, 1969). Rats that have learned to associate a light plus sound cue with a forthcoming small, rather than large, reward, leave an odor trail as a result of that association that, in effect, slows down rats that subsequently traverse the same runway. The odor from rats given a light plus sound cue that signifies a large, rather than small, reward, does not have this effect (Ludvigson et al., 1985).

Brechbühl et al. (2008) present evidence that airborne agents emitted by stressed mice are detected by the Grüneberg ganglion (GG), a chemosensory structure located on the anterior superior nasal septum of the mouse and some other mammals (Grüneberg, 1973). This structure contains several hundred cells, each of which sends an axon to a circumscribed region of the olfactory bulb via a path located along the top of the nasal cavity. Degeneration of this structure secondary to axonotomy resulted in mice that failed to freeze in the presence of vapors from stressed mice. This research strongly implicated the GG in mediating responses to such vapors and found no other stimulants that activated the GG. Using calcium imaging, GG cells were responsive to perfusions of oxycarbonated artificial cerebrospinal fluid containing materials collected from the vicinity of stressed mice, but not to mouse milk, mammary secretions, or a mixture containing odorants and "known isolated mouse pheromones." No agent or agents responsible for the activation of the GG cells that could be construed as a pheromone was identified.

The aforementioned studies indicate that a number of animals can detect, if you will, the emotional state of conspecifics by excreted chemicals. However, the utility of the pheromone concept in explaining such responses seems questionable, and few authors have actually used the term *pheromone* to describe the effect. As with the vast majority of so-called pheromones, no chemical agent or agents have been isolated that produce the response or responses (Brechbühl et al., 2008), although those who have attempted to isolate such agents have concurred that many different compounds are likely involved. One group identified a broad range of cyclic (e.g., terpenes, phenol-derived and furan-derived chemicals) and acyclic compounds (alkanes, alkenes, alcohols, aldehydes, ketones, esters, and acids) in air believed to contain rat alarm substances (Hauser et al., 2005), implying that chemical complexity is likely. The role of the recipient's experience or learning in producing the effects is unclear, although rats can be trained to distinguish between stress-related and non-stress-related odors in operant conditioning tasks (Valenta and Rigby, 1968) and even "artificial" odors are easily conditioned to emotional responses (Cousens and Otto, 1998; Otto et al., 1997). Thus, an odor that has been paired with lithium chloride–induced sickness comes to elicit freezing and analgesia (Richardson and McNally, 2003). The presence of other group members can attenuate alarm responses, although this may require the companions to have been previously habituated to the aversive stimulus (Epley, 1974; Boissy et al., 1998). It is noteworthy that, in all studies examining rodent responses to stress-related odors, the animals were group-reared, providing a potential opportunity for learning freezing and other socially appropriate behavioral responses in interactions where heightened emotions are evident.

In this chapter, classic studies of "releasing" (or "behavioral," "informer," or "signaling") pheromones were reviewed. It is apparent that such agents are rarely, if ever, associated with invariant induction of behavioral or endocrine responses, and that conspecific responses to them depend in large part on the context in which they are presented, as well as the experience and bodily state of the receiving organism. So-called releasing pheromonal effects are nearly always learned or have a major experiential component. Diet markedly alters many putative pheromones. When such agents have been chemically isolated, they typically fail to completely mirror the responses observed for their parent secretions, and multiple chemicals from the same secretions are often capable of producing similar results. In such cases, it would seem impossible to identify "the pheromone."

SIX

The Elusive Snarks

Case Studies of Nonhuman Mammalian "Priming" Pheromones

To seek it with thimbles, to seek it with care;
To pursue it with forks and hope;
To threaten its life with a railway-share;
To charm it with smiles and soap!

For the Snark's a peculiar creature, that won't
Be caught in a commonplace way.
Do all that you know, and try all that you don't:
Not a chance must be wasted to-day!

For England expects—I forbear to proceed:
'Tis a maxim tremendous, but trite:
And you'd best be unpacking the things that you need
To rig yourselves out for the fight.

From the Fourth Fit, "The Hunting," in Lewis Carroll's
The Hunting of the Snark

The influences of urine or body odors on the endocrine state of mammals of the same species have been attributed to "priming" pheromones. Although most claims of such pheromones have been made for rodents, some have been made for non-rodents, including dogs, sheep, goats, and humans. The possibility that humans have priming pheromones has captured the imagination of scientists and laypersons alike, and is among the topics addressed in Chapter 7.

There are few claims for the chemical identification of most so-called priming pheromones and, as in the case of putative releasing pheromones described in Chapter 5, the isolated agents appear to fall short of mimicking the parent stimuli. Moreover, few have stood the test of replication. Although one cannot deny the influences of bodily secretions and excretions on the reproductive physiology of a range of animals, the question is whether such influences can be reasonably and uniquely ascribed to pheromones. To what degree is experience involved? Are the stimuli too complex or idiosyncratic to be identified? To what degree do the stimuli represent abnormal chemical stressors? Can subsets of multi-chemical mixtures induce equivalent responses? All of these questions fall to the heart of the issue as to whether pheromones are involved.

The Ram Sexual Pheromone

In many breeds of domestic sheep, seasonally anestrous ewes isolated from rams for a period of time exhibit, following the presentation of a ram, a rapid rise in luteinizing hormone (LH) levels and a preovulatory LH surge resulting in ovulation 35 to 40 hours later. This effect has been attributed to a pheromone since ovulation can be induced by the presentation of a ram's fleece or extracts of a ram's wool (Knight and Lynch, 1980) and since physical and visual contact are unnecessary to produce this effect (Watson and Radford, 1960).

Cohen-Tannoudji et al. (1994) sought to chemically isolate the "ram sexual pheromone." *Sexually experienced* multiparous ewes served as subjects. Various stimuli were initially assessed for their ability to produce the LH response in the anestrous ewes: (1) male urine, (2) an extract from whole fleece from a sexually experienced ram, (3) an extract from wool collected from either the head or shoulder of such a ram, (4) an extract collected from the flanks and hindquarters of a sexually experienced ram, (5) an extract from female wool, and (6) secretions from the orbital glands below the ram's eyes. Significant LH pulses were induced by (2), (3), and (4), but not by (1), (5), or (6). From these findings, these authors concluded that the active agent is "produced and/or spread over the whole body of the male" (p. 960). However, no specific chemical fraction could be isolated that produced the effect. These investigators candidly noted the following:

The male-female chemical communication involved in primer pheromone action or "male effect" appears to involve a complex olfactory pattern, rather than an accurate balance of a few specific compounds. Thus, the transmission of the characteristics of the male of the species results from a cognitive procedure, and not from a codified message conveyed by some kind of "labeled line." The male's chemical signature could result from a complex mixture of the variety of secretions from the multiple androgen sensitive glands. In such a perspective, the identification of some of the critical compounds is an important step, but does not allow a complete understanding of the message. (p. 960)

It is important to point out that a number of sensory stimuli other than odors can similarly produce this effect. Thus, total ablation of the olfactory bulbs, which typically eliminates input from both the olfactory bulb and vomeronasal organ (via damage to the vomeronasal nerve), does not eliminate the response induced by the presentation of a non-castrate ram. Hence, "other sensory cues may replace the pheromone to trigger LH release" (Cohen-Tannoudji et al., 1986, p. 921). It is also important to note that an extract of male goat's hair and its acid fraction also stimulate such secretion in ewes, violating the pheromone concept of species specificity (Over et al., 1990).

More recently, Gelez et al. (2004a) demonstrated the key role that sexual experience plays in this phenomenon. The proportion of females exhibiting an LH response to male odor was higher in sexually experienced than in non-sexually experienced ewes, and sexual experience influenced the response to the male odor, but not to the male himself. Lavender odor, previously associated with the male, was by itself effective in inducing the LH response. No such effect occurred when the odor was not paired with a male. These researchers concluded, "Our results show that a neutral conditioned stimulus can mimic the effect of the ram odor, suggesting that the LH response may be due to associative olfactory learning. It is possible that ewes have to learn the association between their sexual partner and its odor to assign a 'meaning' to the male chemosignal" (p. 559). These authors go on to state,

Our results show that the ram odor does not possess innate attractive properties, since naïve young ewes displayed similar olfactory investigations of both male and female fleeces. However, the naïve adult ewes smelled the male fleece significantly more than the female fleece. This result may be explained by the past experience of the females. The naïve adult ewes were housed with several groups of conspecifics and had been in contact with the odors of numerous different females. Thus, they had a greater "social experience" than the young naïve ewes that only encountered the females of their living group. (p. 560)

Female Mouse Urine Pheromone That Induces Luteinizing Hormone Release in Male Mice

Coquelin et al. (1984) reported that a pheromone exists in female mouse urine that acts via the vomeronasal organ of conspecific males to "reflexively" release LH. Singer et al. (1988) sought to determine the chemical identity of this putative pheromone. Initially they discovered that a significant amount of biological activity was retained by the urine following dialysis, implying the activity is associated with a urinary protein. Subsequent analysis suggested that the protein probably serves as a "pheromone carrier," since the biological activity disappeared following adsorption chromatography and assay of gel permeation chromatography fractions, before and after degradation of the urinary proteins with proteolytic enzymes (implying the protein is not necessary for the male response in the bioassay). They concluded that the pheromone is not a peptide, as it resisted vigorous proteolytic enzyme treatment. High biological activity, indistinguishable from that of the unfractionated urine, was isolated in a protein-depleted, presumably low molecular weight, fraction containing compounds that are retarded by adsorption on Sephadex.

In fact, no single chemical or small set of chemicals arose from this study that could be labeled a pheromone. It is noteworthy that these mice were sexually experienced and that non-olfactory cues from behaving females also induced the LH release. Thus, this phenomenon could be learned and not specific to chemical cues, although removal of the vomeronasal organ (VNO) did depress the effect. It is worth mentioning that urine odors from predators can alter male reproductive function in a number of species, for example Campbell's hamsters, implying that LH secretion may be influenced by a range of different types of odors (Vasilieva et al., 2000).

Perhaps the ultimate claim for a priming pheromone is that of the blocking of implantation of a recently inseminated female by a strange male. In 1959, the endocrinologist Hilda Bruce reported that only 29% of a group of recently inseminated albino mice became pregnant when paired immediately after insemination with a non-stud male mouse of the wild type strain, compared to 100% that were similarly paired with the stud male (Bruce, 1959). Housing with a non-stud albino male mouse decreased the pregnancy rate to 72%. The latter decrement was present regardless of whether the male albino mouse was intact or castrated (Bruce, 1960a), a finding now believed to be aberrant, given numerous subsequent reports—including ones from Bruce's own laboratory—that castrates are ineffectual in blocking pregnancies (e.g., Bruce, 1965; Spironello-Vella and deCatanzaro, 2001).

A number of studies have verified the Bruce effect, finding in some cases that same-strain ("strange") male mice block pregnancies in 25 to 30% of the females, whereas different-strain ("alien") males do so in up to 80% of the females (Parkes and Bruce, 1962). This general phenomenon has been described in numerous laboratory strains of house mice (*Mus musculus*) (e.g., C3H, BALB/c, CBA), as well as in deer mice (*Peromyscus maniculatus*) and a number of voles (*Microtus agrestis, M. ochrogaster, and M. pennsylvanicus*) (Clulow and Clarke, 1968; Terman, 1969; Stehn and Richmond, 1975; Bronson et al., 1969; Clulow and Langford, 1964; Bronson and Eleftheriou, 1963; Bruce, 1959, 1960a; Watson et al., 1983; Storey, 1986). It apparently does not occur in all *Mus* strains (Kakihana et al., 1974; Bruce, 1968) or in gerbils (*Meriones unguiculatus*) (Norris and Adams, 1979). The female appears to be most vulnerable to pregnancy block within 48 hours of coitus, with exposure during the first 12 hours being sufficient in the majority of cases (Bruce, 1961). The effectiveness of the pregnancy block is reportedly not augmented by increasing the number of males to which a female is exposed (Chipman and Fox, 1966b; Bruce, 1963; Bronson and Eleftheriou, 1963), although one study found that six strange males blocked a greater percentage of pregnancies than a single male (85% versus 42%), conceivably reflecting the intensity or complexity of the stimulus (Chipman and Fox, 1966a).

The Bruce effect has been attributed to a pheromone, since (1) it occurs even when the strange or alien male is separated from the female by a wire partition (Bruce, 1959); (2) it is eliminated by olfactory bulbectomy or damage to the female's accessory olfactory system (Bruce and Parrott, 1960; Rajendren and Dominic, 1985; Bellringer et al., 1980; Reynolds and Keverne, 1979; Lloyd-Thomas and Keverne, 1982); (3) it can be blocked by anointing the strange male with a masking odor, that is, a perfume (Sahu and Dominic, 1983); and (4) urine or soiled bedding from a strange male is as effective, in many strains, as the strange or alien male himself in producing the effect (Dominic, 1965, 1964, 1966b). Nonetheless, at least limited physical contact with the male seems to be needed in some strains or test situations to block pregnancy (deCatanzaro et al., 1995b). While in one study exposure of recently mated Parks albino females to urine-soiled bedding from strange males in boxes was ineffectual (Bruce, 1960b), the pregnancy block could be produced by housing the mice in tall, poorly ventilated, glass jars containing bedding to which strange male odor urine had been added. Another study using the same type of albino mice found that once-daily renewal of the soiled bedding is markedly inferior to twice-daily renewal, suggesting that "the operative substances are evanescent, highly volatile or highly labile, probably both" (Parkes and Bruce, 1962, p. 307). Three 15-minute exposures over a 4-day test period have been found sufficient to produce the phenomenon in outbred Swiss strain females exposed to wild type males (Chipman et al., 1966). There is evidence that the urinary chemicals responsible for mediating the Bruce effect may depend upon MHC class 1 peptides (Kelliher et al., 2006). The accessory sexual glands (e.g., the vesicular and coagulating glands) or the preputial glands are unlikely important, since pregnancy blocking efficacy remains in mice lacking these organs (Zacharias et al., 2000; Hoppe, 1975). Administration of epiandrosterone, androstenedione, androsterone, or testosterone to SJL female mice results in their ability to block pregnancies; administration of progesterone or dehydroepiandrosterone does not (Hoppe, 1975).

The physiological basis for the Bruce effect is complex. An intact VNO seems necessary for the induction of the Bruce effect (Kelliher et al., 2006), even though the effect is not diminished in transgenic mice lacking the transient receptor potential channel TRPC2, a key element of VNO signal transduction (Kelliher et al., 2006). TRPC2$^{2/2}$ mice have been widely used as a genetic model for investigating general VNO function (Brennan and Keverne, 2004). Memory of the stud male via the vomeronasal organ appears to be important, as the effect is eliminated by infusion of phentoamine or other memory-blocking agents into the female's accessory olfactory bulb after initial mating (Kaba and Keverne, 1988; Kaba et al., 1989). However, the recognition of the strange versus familiar male depends upon an intact noradrenergic projection to the olfactory bulbs (Keverne and de la Riva, 1982). Importantly, factors that alter leuteotrophic function are intimately involved. Thus, the Bruce effect is (1) blocked by injecting recently inseminated females with the primary leuteogenic agent in the mouse, prolactin, or with progesterone during the strange male odor exposure period (Bruce and Parkes, 1960; Dominic, 1966b; Rajendren and Dominic, 1987; Helmreich, 1960); (2) blocked by the implantation of a functional ectopic pituitary graft known to produce prolactin (Dominic, 1966a, 1967, 1966b; Bronson et al., 1969); (3) absent in postpartum pregnant females whose prolactin production is induced by suckling (Bruce and Parkes, 1961); and (4) blocked by the injection of reserpine (Dominic, 1966b, 1966c), an agent that depletes stores of catecholamines and serotonin in the brain and suppresses the inhibitory hypothalamic center that controls release of prolactin (Dominic, 1966b). Since pregnancy block can occur in some females exposed to alien males for as little as 12 hours (Bruce, 1960a), it has been suggested that LH release may be an initial determinant of the pregnancy blockage (Chapman et al., 1970). Estrogen may also be involved, since (1) the administration of estrogen—including minute amounts applied to the nose—eliminates successful implantation in non-lactating female mice (Bloch, 1971; Brennan et al., 1990), (2) the presence of males appears to enhance the synthesis and release of FSH in gonadectomized female mice (Bronson and Desjardins, 1969), and (3) antibodies to 17β-estradiol prevent the Bruce effect in recently inseminated females (deCatanzaro et al., 1995a). Exogenous administration of androstenedione and dehydroepiandrosterone, androgens that can be converted to estrogens, is capable of blocking a recently inseminated female's pregnancy; the major stress-related adrenal hormone corticosterone, which is not convertible into estrogens, fails to do so (de Catanzaro et al., 1991). Epinephrine is ineffectual, even though it does disrupt female mating behavior (deCatanzaro and Graham, 1992). Estrogen's influences are likely not independent of those of prolactin, however, since estrogen activates prolactin release, a phenomenon dependent upon a functional adrenal gland (Milenkovic et al., 1986).

The role of the adrenal gland in the Bruce effect appears to be strain dependent. In some strains, adrenalectomy has little effect on strange-male induced pregnancy block (Sahu and Dominic, 1981), whereas in others attenuated adrenal responses or adrenalectomy mitigate the effect (Marsden and Bronson, 1965; Snyder and Taggart, 1967). Exogenous adrenocorticotropic hormone (ACTH) is capable of inhibiting luteinization in adrenalectomized mice, but is more effective in mice with adrenal glands (Christian et al., 1965). Such effects may not be independent of estrogen, however, as injections of ACTH can produce increases in estrogen levels in some species (Strott et al., 1975; Arai et al., 1972).

Regardless of the hormonal factors involved and the possibility of minimal influences of corticosterone, it would seem likely that responses to acute stress are involved. As noted in several reviews (e.g., deCatanzaro and MacNiven, 1992; Marchlewska-Koj, 1997), blockage of implantation is not uncommon in rodents and can be induced by a variety of stressors, including brief periods of starvation. In mice, the pregnancy of recently inseminated females can be blocked by (1) human handling (an effect eliminated by the injection of progesterone) (Runner, 1959; Weir and DeFries, 1963; Chipman and Fox, 1966a), (2) enforced swimming for 3 minutes and/or exposure to loud tones and open areas (Weir and DeFries, 1963), (3) exposure to male rats (with or without tactile contact) (deCatanzaro, 1988), (4) exposure to predators (deCatanzaro, 1988), (5) exposure to male or female rat urine (deCatanzaro, 1988), and (6) the induction of nutritional or restraint stress (McClure, 1959; Euker and Riegle, 1973), the latter of which is reduced by the administration of estrogen antibodies during the pre-implantation and early implantation stages (deCatanzaro et al., 1994). The effectiveness of the Bruce effect, when mating is involved, is related to the aggressiveness and sexual behavior of the strange mouse, with intromissions being particularly important (deCatanzaro and Storey, 1989; Storey and Snow, 1990). In the deer mouse *Peromyscus maniculatus,* the nature of the post-insemination environment influences implantation (Eleftheriou et al., 1962), such that post-insemination housing in different-sized cages decreases pregnancy success by 30 to 60%, depending upon the difference in size of the new environment relative to that of the old one. In rats, exposure to noxious sounds four to six days after mating reduces the number of pregnancies, as does chronic restraint (Zondek and Tamari, 1967; Euker and Riegle, 1973).[1] Endogenous estrogen rises in response to acute stress during early pregnancy as well as during non-pregnant states (MacNiven et al., 1992; Shors et al., 1999), adding further credence to the notion that estrogen may be a primary factor in stress-induced blockage of implantation.[2]

Evidence that stress-reducing manipulations can protect against the Bruce effect includes observations that (1) the presence of the stud along with a strange male mitigates the strange male's blocking ability to some degree (Parkes and Bruce, 1961; Terman, 1969), possibly by decreasing the strangeness of the situation or counteracting the strange male odor; (2) replacement of a strange male with an original stud that was present during the period of the previous pregnancy prevents the pregnancy block, whereas replacement with a stud that had limited contact with the female does not (Bloch, 1974);[3] (3) familiarization of the female with a male before she is mated with another male mitigates the familiarized male's ability to block the pregnancy, regardless of the strain of the female, the strain of the stud male, or the strain of the familiarized male (Furudate and Nakano, 1981; Parkes and Bruce, 1961);[4] and (4) the presence of other females or their urine—even urine from spayed or androgenized females—during the critical post-coital period can effectively prevent the pregnancy block in some strains, with the prophylactic effect being directly related to the number of females present (at least up to the largest number tested, i.e., 6) (Bruce, 1963; Dominic, 1965). Whether this reflects the calming effect of such odors, familiarity with such odors, belief that other females are present, prior conditioning induced in the suckling setting, or some type of chemical cross-adaptation to the strange male odor is not clear. In laboratory rats, stress is reduced proportionate to the proximity of a second rat that is also undergoing stress (Leshem and Sherman, 2006). In an atypically reactive house mouse strain, in which the threshold for pregnancy blockage is low, repeated handling of the pups in infancy decreases their susceptibility to the Bruce effect in later life (Bruce et al., 1968), presumably reflecting mitigation of later stress responses (King, 1959; Denenberg et al., 1977; Levine and Broadhurst, 1963). It is noteworthy that hyperprolactinemia is associated with suppression of stressor-induced elevations in plasma corticosterone levels (Drago et al., 1986; Endroczi and Nyakas, 1974), and that reserpine reverses the density-dependent adrenal hypertrophy and reproductive decrements observed in male house mice (Christian, 1956).

It is of particular interest that most, if not all, drugs and hormones that have been shown to block the Bruce effect, including amitriptyline, chloropromazine, haloperidol, pimozide, progesterone, prolactin, propranolol, and reserpine, with the exception of those targeted on odor memory formation of the stud or strange male (e.g., phentolamine, anisomycin), have anxiolytic or antidepressant properties (Torner et al., 2001; Rajendren and Dutta, 1988; Saletu et al., 1975; Bloch and Wyss, 1973; Sahu and Dominic, 1980; Dominic, 1966b, 1966c). Whether the perfumes employed in the study by Sahu and Dominic (1983) that eliminated the effectiveness of strange males in blocking the pregnancy of recently inseminated females did so by decreasing the stress associated with the strange male odor, rather than simply masking the strange male odor, requires further study. Such antidepressants as fluoxetine, amitriptyline, desipramine, and buspirone enhance habituation to novel stimuli in olfactory bulbectomized rats—rats that exhibit many symptoms of being stressed. This raises the possibility that such agents may depress the degree of novelty of a strange male or his odors on implantation (Mar et al., 2000). Of course, such actions need not be independent of the influences of these agents on the hypothalamic-pituitary-gonadal axis. Interestingly, in humans some pleasant odors reportedly reduce stress responses (Ludvigson and Rottman, 1989).

If one accepts the proposition that the Bruce effect is induced by stress, then it would seem that testosterone, or some other testicular-influenced agent, is involved in the production of the stress-inducing odor or agent, since (1) urine from male castrates is typically ineffectual (Bruce, 1965; Spironello-Vella and deCatanzaro, 2001); (2) males are capable of blocking pregnancy only after the age of puberty (Bruce, 1965); (3) urine from males housed alone is less effective than urine from males housed near females (whose testosterone titer would be expected to be comparatively higher) (de Catanzaro et al., 1999); (4) the anti-androgen cyproterone acetate mitigates the pregnancy-blocking ability of the male's urine (Bloch and Wyss, 1973); and (5) male mice that have achieved sexual satiety are less effective than those that have not in producing the pregnancy blockage (Spironello and deCatanzaro, 1999), presumably reflecting decreased testosterone titer (Batty, 1978). It is possible that testosterone simply produces a strong salient smell that, without familiarization or adaptation on the part of the female, is (1) clearly discernible to the female, (2) allows for accurate differentiation between different males, and (3) elicits enhanced sensorineural activity. Urine from non-castrate male rodents is much more intense, even to humans, than that from castrates or females.

Whatever the role of testosterone, it seems clear that the female rapidly learns the odor of the stud male and by further exposure accommodates or adapts to the stud's odor over time, making it either weak or familiar, and relatively non-stressful.[5] Presumably the degree of strangeness of the strange male is a function of the degree of qualitative difference between its odor and that of the stud—a difference that must be learned and somehow discerned via the accessory olfactory system (Brennan and Keverne, 1997). If castrate males have relatively little smell, then they or their odors would probably not seem particularly strange or stressful to the female. Unfortunately, no control studies have been performed that have attempted to equate artificial odors on intensity or quality with non-castrate male urine odors, or to differentially odorize the stud and strange males to see if such odors can produce the Bruce effect.

There have been few attempts to identify the putative pheromones responsible in the male urine for inducing pregnancy blockage. Monder et al. (1978) exposed, in two studies, 139 recently inseminated DBA female mice to various volatiles, including water, SJL/J male mouse urine, a dichloromethane lipid exact of the urine, or dichloromethane alone for a 5-day period beginning 24 hours after the discovery of a vaginal plug or sperm in the vagina. Ten days after insemination the uteri were checked for signs of pregnancy and the number of fetuses counted. The females exposed to the urine had significantly fewer pregnancies than the females exposed to the other substances, although there was a non-significant trend for fewer pregnancies in the group exposed to the dichloromethane lipid extract.

Inherently, the Bruce effect would seem not to be well suited as being due to a pheromone, since the pheromone would have to vary from male mouse to male mouse, so long as the mouse differs from the stud male in its identifying chemical composition or odor. Hence, there could be as many pheromones as there are male mice that can be identified as individuals, at least as individuals that have testosterone-related odors. Moreover, this phenomenon has a major learning component associated with it, and seems to be a subclass of a group of phenomena that relies upon the fragile nature of the implantation process. Where is the pheromone in this situation? Certainly, no pheromone has been chemically identified.

The Puberty-accelerating Pheromone of Male Mice (Vandenbergh Effect)

Laboratory studies have shown that prepubertal female mice housed or otherwise exposed to non-castrate males or their urine attain puberty sooner than females not so exposed, as measured by the time of vaginal opening, the time of first estrus, and uterine weight (Andervont, 1944; Castro, 1967; Vandenbergh, 1969, 1967; Cowley and Wise, 1972; Novotny et al., 1999a). Such acceleration of puberty, which has been classically considered as being due to a priming pheromone, is stronger when the male is present, conceivably reflecting the addition of contact stimulation (Bronson and Maruniak, 1975; Drickamer, 1974, 1975). Application of male urine directly to the oral-nasal grooves of post-weanling female mice accelerates the time of onset of first estrus by 4 to 6 days relative to controls exposed to tap water. Exposure for 3 days between the ages of 21 and 29 days is sufficient to advance puberty, and intermittent stimulation can be superior to continuous stimulation (Drickamer, 1987). Exposure to male urine tends to shorten the interval between the day of vaginal opening and the day of first estrus, implying that the urine accelerates sexual maturation rather than just triggering puberty (Colby and Vandenberg, 1974). Urine from males whose preputial glands are removed is as effective as urine from intact males.

It is now known that (1) urine from dominant males produces greater acceleration than urine from subordinate males; (2) the male urine must be present for at least 2 to 3 hours/day, or the male must be present for 1 hour/day, to produce the acceleration; (3) urine from the same male presented each day produces the same degree of acceleration as urine from different males presented each day; (4) urine from the father or a full brother exerts the same degree of acceleration as urine from unrelated males, and (5) excreted or bladder urine from adrenalectomized males is as effective as urine from an intact male (Drickamer, 1986a). However, the presence of an adrenalectomized male does not produce the same degree of acceleration as the presence of an intact male, possibly because adrenalectomized males pursue young females less actively and attempt fewer mounts (Drickamer, 1983). More recent studies have revealed that urine from a number of different types of mice in addition to mature males accelerates the time of puberty, including urine from estrous, pregnant, or lactating female mice (Drickamer, 1984a, b; Cowley and Pewtress, 1986). Indeed, urine from male *rats* also accelerates the time of puberty in female mice, negating the notion of species specificity (Colby and Vandenberg, 1974).[6] Thus, from a functional or evolutionary perspective, the situation is complex and of unknown ecological import.

Numerous attempts have been made to isolate the "puberty-accelerating pheromone" of prepubertal female house mice. A single unique and unitary agent responsible for pubertal acceleration has not been found and both volatile and nonvolatile urinary constituents likely play a role.[7] In a pioneering study, two volatile amines, identified in the urine of male mice of the ICR strain, were found to decrease following castration (Nishimura et al., 1989). A mixture of these amines—isobutylamine and isoamylamine—was reported to accelerate the time of vaginal opening in females of the same strain. However, the amounts of materials used were far higher in concentration than present in mouse urine (0.1 M versus 10^{-8} to 10^{-9} M) and other researchers were unable to replicate these findings (Price and Vandenbergh, 1992).

In 1989, a group headed by Milos Novotny reported that supraphysiological concentrations of a number of volatile ketones found in mouse urine advanced female puberty by approximately three days and extended the period of vaginal cornification in up to 75% of exposed individuals (Jemiolo et al., 1989). A decade later, this same group (Novotny et al., 1999a) isolated from the urine of male mice a "unique urinary constituent" that is "a pheromone that accelerates puberty in female mice." After investigating urinary fractions associated with increasing uterine weight of females, these authors reported that

> the active pheromone was identified as 5,5-dimethyl-2-ethyltetrahydrofuran-2-ol and/or its open-chain tautomer (6-hydroxy-6-methyl-3-heptanone). A series of cyclic vinyl ethers were isolated from chromatographically active fractions of the urine. Because these compounds did not

accelerate puberty, we postulated that these ethers were degradation products of a lactol (5,5-dimethyl-2-ethyltetrahydrofuran-2-ol). The lactol was then detected in the mouse urine extract using a silylation agent. Synthetic 6-hydroxy-6-methyl-3-heptanone had strong biological activity, whereas its close structural analogs did not. (p. 377)

How well did the biological activity of 6-hydroxy-6-methyl-3-heptanone (6-HMH) compare to the urine itself? In the SJL/J × SWR/J animals, the respective mean (± SEM) uterine weights (mg) of those exposed to the whole urine, to 6-HMH, and to water were 80.0 (± 6.8), 64.1 (± 4.9), and 31.2 (± 3.2). These three means differed significantly from one another. In the IRC/Alb mice, these three respective values were 102.9 (± 12.0), 70.3 (± 5.8), and 37.0 (± 2.9). Again, these values differed significantly from one another. Note that in both mouse lines the activity of the 6-HMH was significantly lower than that of the urine alone. The respective percentages of the SJL/J × SWR/J mice exhibiting a uterine response to the male urine, 6-HMH, and water alone were 71%, 57%, and 3%. Corresponding percentages for the IRC/Alb mice were 76%, 59%, and 22%. The responding females were defined for the urine and 6-HMH groups as those whose uterine weights fell outside one standard deviation of the "range of the weights in the control group." Despite this liberal criterion for assessing responsiveness, it is clear that considerably fewer mice responded to the 6-HMH stimulus than to the urine.

In the same year, this group also published a study entitled "Positive Identification of *the* [my emphasis] Puberty-accelerating Pheromone of the House Mouse: The Volatile Ligands Associated with the Major Urinary Protein" (Novotny et al., 1999b). In this work, a number of structurally diverse small ligands, all of which bind to the MUP, were found individually to accelerate puberty in the recipient females, as were some other agents (Novotny et al., 1999b). In contrast to the findings of another group (Mucignat-Caretta et al., 1995a), neither MUP alone, nor an associated hexapeptide, promoted uterine growth. The substances found to significantly increase uterine weight relative to water were 3,4-dehydro-exo-brevicomin (DB), 2-sec-butyl-4,5-dihyrothiazole (BT), MUP from male mice (MM), MUP from female mice (MUP), β-farnesene (β), a mixture of α- and β-farnesene (αβ), lactol (L), and MUP from testosterone-treated females (FTM). It is noteworthy that these authors talk about *the* pheromone involved in puberty acceleration yet delineate multiple agents in addition to 6-HMH.

The aforementioned findings strongly suggest that multiple urinary chemicals, alone or in combination, are involved in pubertal acceleration in female house mice, although other chemicals, such as those from scent glands, may also play a role. As stated by Jemiolo et al. (1989), the ratio of numerous urinary compounds likely determines the "pheromonal response of female mice." That being said, the pre-weaning rearing condition can also influence the degree to which urinary chemicals influence the timing of puberty. For example, Mucignat-Caretta et al. (1995b) reared females under three conditions: with both parents, with two females, or with two females and the presence of urine from adult males. Nine days after weaning, the prepubescent females were exposed daily to either adult male urine or prepubertal male urine. The adult male urine resulted in larger uteri and more cornified vaginal smears than the prepubertal male urine in the two groups reared with male odors (i.e., the one with both parents present and the one with two females and male odor). Interestingly, in the group reared with only females, no significant changes in uterine weight or vaginal smears were observed.

It is important to note that pubertal acceleration is not uniquely determined by chemical stimuli in laboratory mice, as suggested by the observation that less exposure time to the whole mouse is needed than to the odor alone to induce the phenomenon. This is illustrated by a study showing that both olfactory and tactile cues work synergistically to facilitate pubertal onset (Bronson and Maruniak, 1975). In this study, exposure of immature females to urine from intact males for 54 hours resulted in small but consistent uterine growth. Living with a castrated male for the same time period had no influence on uterine growth. However, simultaneous exposure to the intact male urine and living with the castrate male produced uterine growth nearly to the same degree as cohabitation with an intact male. By limiting auditory, visual, and contact stimulation, these investigators found tactile and chemical stimuli to be the main contributors to the pubertal acceleration (Figure 6.1).

Numerous sensory and nutritional factors, not the least of which is the light/dark cycle, alter the time of puberty in a wide variety of mammals. Nonetheless, in mice the presence of an adult male seems to be more important in accelerating the time of puberty than either nutrition or

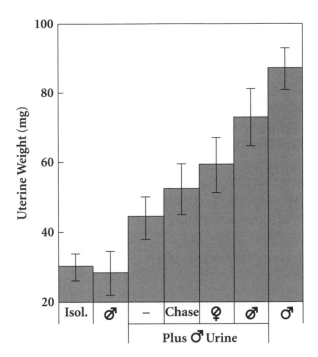

FIGURE 6.1. Mean (± SEM) uterine weight after 54 hours of exposure to isolation, cohabitation with intact or castrate males, or exposure to intact male urine while either isolated or cohabitating with ovariectomized females or castrate males. "Chase" refers to isolated females exposed to intact male urine and periodically chased around their home cages by the gloved hand of an experimenter. n = 14 or 15 females in each treatment. Modified from Bronson and Maruniak (1975).

light/dark cycles (Vandenbergh et al., 1972). Stress, or the mitigation of stress, also appears to influence the timing of rodent sexual maturation, and it is not clear whether some of the urinary chemical stimuli presented to mice in the laboratory are stressors for them.[8] As shown in Figure 6.1, chasing the young female mouse around the cage with the experimenter's glove significantly increases uterine weight. When female rat pups are handled and placed individually in separate containers for 3 minutes each day from birth to 24 days of age, they display first estrus, on average, 10 days before unhandled controls (Morton et al., 1963).[9] When such handled rats are housed in groups of 10 in a large cage with an enriched environment, the age of pubertal opening is delayed (Swanson and van de

Poll, 1983). Intense visual, auditory, and/or electrical stimuli applied shortly before the time of normal physiological puberty accelerates the time of puberty in rats, although if such stimuli are applied much in advance of this time, puberty is delayed (Beardwood, 1982; Árvay and Nagy, 1959; Árvay, 1964; Ameli and De Marini, 1966). Repeated exposure of young rats to cold stress brings about vaginal opening 3 to 4 days earlier than in controls (Mandl and Zuckerman, 1952). In house mice, acute stress, either via tail shock or forced swimming, increases estrogen titer, conceivably providing a mechanism for advancing puberty (Shors et al., 1999). Interestingly, changes in adrenocortical function precede the onset of persistent estrus in rats exposed to constant light before puberty (Ramaley, 1977).

Even in humans, the time of puberty appears to be influenced by stress. Thus, girls from divorced families and families with greater interparental conflict tend to have an earlier menarche than girls from intact and more functional families (Wierson et al., 1993) and prolonged exposure to increased adrenal or ovarian steroids sufficient to advance bone maturation is associated with early pubescence (Boyar et al., 1973). Interestingly, in families in which the biological father is not the mother's romantic partner, dyadic stress accounts for nearly half of the variation in the pubertal timing of daughters. The presence of the stepfather, rather than the absence of the biological father, best accounts for the earlier maturation in girls living apart from their biological fathers (Ellis and Garber, 2000).

In light of the complexity of acceleration of maturation, including its potential multisensory determination and association with stressors that may well include urine-based chemicals, should chemical substances that are found to influence, to varying degrees, the time of pubertal onset be termed *pheromones*? After nearly 40 years since the report of this phenomenon, it would appear that the "puberty-accelerating pheromone" has yet to be found.

The Estrus Suppression Pheromone of Grouped Female Mice (Lee–Boot Effect)

In general, crowding decreases reproduction in a wide range of mammals (Crew and Mirskaia, 1931; Christian and LeMunyan, 1958), as reflected in endocrine changes in both sexes (Terman, 1968; Christian and Davis, 1964). In the case of house mice, grouping females together in laboratory cages without the presence of males results in pseudo-pregnancy (Van der Lee and Boot, 1955, 1956; Dewar, 1959; Ryan and Schwartz, 1977) or, less frequently, prolonged periods of diestrus (Lamond, 1958 a, b; Champlin, 1971; Whitten, 1959; Lamond, 1959). Diestrus is more likely to occur when the mice are housed in large groups (30 per cage), whereas pseudopregnancy is more common in smaller groups (Parkes and Bruce, 1961). This phenomenon has been termed the *Lee–Boot effect* after the two endocrinologists who first described it. In the case of the C57BL/5 and BALB/ci *Mus* strains, as group size increases (test groups: 1, 2, 4, 6, and 8 females/cage), the proportion of females exhibiting 4- or 5-day cycles decreases, with more individuals exhibiting longer and more variable cycles (Champlin, 1971). An asymptote in grouping effectiveness appears to occur in a group size of about four in the C57BL strain, and possibly around six in the BALB/Ci strain (Champlin, 1971). Whitten (1959) reported that (1) the acyclicity often continues for months after grouping; (2) grouping is associated with lighter-weight ovaries; (3) the acyclicity is not influenced by blinding of the grouped animals; (4) a few animals continue to exhibit regular cycles with seemingly prolonged estrous phases; and (5) the vaginal smears of grouped animals are typically mucified, reminiscent of smears from animals given estrogen at doses too low to elicit vaginal cornification.

The Lee–Boot effect is said to be due to a pheromone, since (1) it is eliminated by olfactory bulbectomy or vomeronasal organ excision (Reynolds and Keverne, 1979) and (2) excreta from crowded females induces analogous changes in the estrous cycles of singly housed females (Champlin, 1971). An intact ovary is seemingly a necessary element in the production of the chemical stimulus, since neither grouped ovariectomized female mice nor their urine produce the effect. The effectiveness of the urine is restored following estrogen and/or progesterone injection of the donors (Bloch, 1976; Clee et al., 1975). Since grouped adrenalectomized females also fail to produce urine that induces acyclicity, the adrenal gland may be involved in the production of the stimulus (Ma et al., 1998).

One explanation of the Lee–Boot effect is that it represents stress responses associated with the recognition of abnormal grouping of female mice that would rarely if ever be found in nature. Such grouping results in an increase of prolactin, a hormone associated with stress in mice which, in turn, induces the pseudopregnancy. Thus, treatment with a drug that lowers prolactin levels (CB 154) overcomes the estrus suppression 72 hours after its administration, and treatment with haloperidol, a drug that raises plasma prolactin, induces estrus suppression (Reynolds and Keverne, 1979). Grouped intact diestrous female mice, as well as grouped ovariectomized female mice with silastic implants producing low levels of estrogen, have prolactin levels four times greater than those observed in isolated females (Bronson, 1976b). Adrenalectomized mice of at least some strains fail to show the Lee–Boot effect (Ma et al., 1998), a phenomenon that would be expected if stress was producing the effect, since such mice exhibit a markedly attenuated stress-induced secretion of prolactin (Milenkovic et al., 1986). Corticosterone repletion restores the estrus suppression (Ma et al., 1998), again implicating stress as a key element of its genesis. Olfactory bulbectomy would be expected to reduce the ability of the mice to recognize the group composition, whether represented by the mice themselves or by their odors, thereby mitigating the stress response. It should be noted, however, that bulbectomy has massive influences on a range of behavioral, hormonal, neurochemical, and anatomic systems. This includes degeneration of some brain structures and altered cholinergic function (Brunjes, 1992; Kelly et al., 1997), although serum glucocorticoids are not altered (Williams et al., 1992).

From this perspective, the Lee–Boot effect is not much different from other stress-related responses of rodents. Acute stress, such as that induced by changes in the social or physical environment (e.g., Brown et al., 1977), markedly increases prolactin and can thereby induce pseudopregnancy and other endocrine changes. Exposing cycle-stable female rats every hour to bells and flashing spotlights for five minutes results in a transient lengthening of estrous cycles and then to a permanent stage of diestrus, interrupted by some sporadic days of estrus (Árvay and Jasmin, 1964).[10] Exposure of female rats to intense auditory stimulation for five minutes a day over a number of days markedly increases adrenal weight and decreases the number of ripening ovarian follicles (Jurtshuk et al., 1959; Sackler et al., 1959). Electric foot shock or the stress of a surgical operation depress cycling in a similar manner (Marchlewska-Koj et al., 1994; Swingle et al., 1950), although social stress may operate via a different mechanism than stress induced by electric shock. Thus, unlike shocked mice, grouped female mice have increased plasma progesterone and well-developed corpora lutea (Marchlewska-Koj et al., 1994). Administration of reserpine, an anxyiolitic agent, prevents the adrenal enlargements and related stress-induced changes in other organs (Christian, 1959). The stress-induced increase in serum prolactin induced in rats by acute stress is decreased by dexamethasone administration (Euker et al., 1975).

It is important to point out that group-housed female mice not only have higher plasma corticosterone levels than singly housed mice (Nichols and Chevins, 1981; Brain and Nowell, 1971; Solem, 1966), but also have higher corticosterone levels than those found in mixed-sex or all-male groups (Christian, 1960; Mody and Christian, 1962). This effect varies with the time of day, being noticeable before, but not after, the onset of the dark phase of a 12:12 L:D cycle (Nichols and Chevins, 1981). Interestingly, females whose fertility has been inhibited by confined crowding become fertile immediately after they are allowed to disperse into a large pen (Crowcroft and Rowe, 1958).

Since it is unlikely that sizable groups of female rodents exist in the absence of males in wild habitats, it has been suggested that the Lee–Boot effect is a laboratory artifact (Bronson and Coquelin, 1980).[11] Moreover, the generalizability of the Lee–Boot effect to species other than *Mus musculus* is not clear. In virgin female Wistar rats (*Rattus norvegicus*), grouping females together without males from the time of weaning results in a shortening of estrous cycles from five to four days in some animals, but not in anestrus or pseudopregnancy. In deer mice (*Peromyscus maniculatus*), exposure to urine from grouped females has no significant influence on ovarian or uterine growth relative to distilled water (Terman, 1968). In gerbils (*Meriones unguiculatus*), inhibition of sexual maturation is caused by exposure to several *familiar* conspecifics, regardless of their sex, conceivably reflecting inbreeding avoidance (Clark and Galef, 2002). In golden hamsters (*Mesocricetus aruatus*), grouping females together without males does not influence estrous cyclicity, but does influence the number of females exhibiting lordosis, as well as normal mating behavior. This effect apparently depends upon contact stimuli, but not olfaction, since it is mitigated by eliminating contact between the animals, but not by olfactory bulbectomy, which would eliminate input from both the main and accessory olfactory systems (Brown and Lisk, 1978).

In light of the aforementioned findings, it is perhaps not surprising that no one has ever found a chemical or a small set of chemicals that can explain the Lee–Boot effect. If, in fact, this effect reflects a stress response based upon the ability of female mice to discern, using olfaction, the relative number and sex of female conspecifics in the immediate vicinity, then the search for the involved pheromone clearly represents a snark hunt.

The Estrus Induction Pheromone of Mice (Whitten Effect)

Whitten (1959) demonstrated that the majority of female mice housed together in groups of 30 exhibited estrus 4 days after either being placed individually with a male partner or after the placement of a wire basket containing a male mouse into their cage. The latter observation suggested that this phenomenon, now known as the Whitten effect, is mediated by a pheromone, and the Whitten effect is now considered to be a classical example of an endocrinological change induced by a primer pheromone.

While males or their odors can "release" the estrous cycles of females from group-induced suppression, it is of interest that Whitten noted in his classic paper that when previously grouped females were housed individually in single cages, they promptly returned to normal estrous cycles that did not differ from those of ungrouped females (Whitten, 1959). He stated, "There was no significant difference between the two groups, which shows that oestrous cycles returned promptly after segregation and were suppressed shortly after grouping" (p. 103). This observation, as well as a number of others, suggests that significant change in a group-housed female mouse's environment can reactivate the reproductive endocrine system. From this perspective, salient male odors are but one set of stimuli that can result in such reactivation. Indeed, even cross-species influences have been noted in terms of altering the frequency of reproductive acyclicity. Thus, housing female rats next to female hamsters (with a wire partition) results in fewer female rats exhibiting acyclic cycles than housing female rats next to female rats, although more six-day cycles are observed in rats reared near hamsters (Weizenbaum et al., 1977).

That being said, one must concede that the factors involved in such reactivation are poorly defined, although acute activation of the sympathetic nervous system seems likely to be involved. As noted by Marchlewska-Koj and Zacharczuk-Kakietek (1990), exposure of estrous female mice to the odors of male bedding leads to a higher level of acute corticosterone release than that induced by exposure of such mice to the odors of grouped female mice or to the stress of being moved to a new clean cage. On a similar note, it is of interest that the estrous cycles of old non-cyclic rats can be brought back by the administration of progesterone, ACTH, ether stress, or L-Dopa, with ACTH and progesterone being the most effective in inducing regular cycles (Huang et al., 1976). In this study, most of the old rats returned to constant estrus or irregular cycling after discontinuance of the treatments.

When chemical stimuli are involved, those responsible for the reinduction of cycling are generally found in the urine and depend, like those involved with the Bruce effect, upon a non-castrated male for efficacy. While urine alone produces this effect, preputial gland secretions may contribute to the post-grouping acceleration or synchronization of estrus. Thus, urine from gonadally intact preputialectomized males is less effective in accelerating estrus than urine from gonadally intact non-preputialectomized males (Chipman and Albrecht, 1974). Preputial gland secretions are complex, with more than 50 identified volatile components having been isolated that are influenced by such factors as the mouse's housing condition and social rank (Pohorecky et al., 2008). Nevertheless, a mixture of two major volatiles from the mouse preputial gland, E,E-α-farnesene and E-β-farnesene, has been identified that mimics, in sexually experienced females, the estrus-induction capabilities of the whole preputial homogenate obtained from intact males, implying "estrus induction signals" can come from multiple sources (Ma et al., 1999).

As with other so-called priming pheromones, strain differences are present. Bartke and Wolff (1966) reported that exposure to heterozygous A^y/a male mice did not elicit the Whitten effect in females, but that exposure to a/a males did so. Interestingly, the urine from *both* of these types of mice induce the Bruce effect, implying that if a pheromone is involved in the Whitten effect, it must differ from the one that causes pregnancy block (Kakihana et al., 1974). Species differences are also present. For example, even though irregular and lengthy estrous cycles analogous to those seen in grouped female mice can be induced by underfeeding rats, the presence of male rats does not reactivate normal cycles. Indeed, such presence lengthens the cycles and increases the frequency of anestrus

(Purvis et al., 1971; Cooper et al., 1972; Marchlewska-Koj, 1983), although the lengthened cycles can be decreased by presenting different males each day for 15 hours into the female rat's cage (Cooper and Haynes, 1967). Physical contact is not required, as males that are separated from the females by wire mesh induce this shortening of the female cycles (McNeilly et al., 1970).

Estrous Synchrony Pheromones of Rats

In contrast to the Lee–Boot effect in mice, group housing of female rats has been claimed by one group of investigators to induce estrous synchrony and more regular estrous cycles (McClintock, 1978). This phenomenon was attributed to pheromones, since air circulated among the cages was found to induce the effect. Like most of the other studies described this chapter, no chemical or set of chemicals has been identified that produces the phenomenon.

In the first experiment of the defining study (McClintock, 1978), female rats housed in groups were reported to have begun synchronizing their estrous cycles within a relatively short period of time (i.e., after three to four cycles). In this experiment, 30 females that exhibited regular 4-day cycles were divided into 2 groups. The first of these 2 groups was subdivided into 3 groups of 5 animals apiece. The 15 rats of the second group were housed individually. For the next 30 days, vaginal smears were recorded daily. After this period, the housing conditions of the groups were reversed, and vaginal smears were again recorded daily for 30 days. Two control groups were formed for both the first and second 30-day recording periods. The first was constructed by randomly assigning the vaginal smear records of the 15 isolated females to 3 groups of 5 each, and the second was constructed by randomly and similarly assigning the vaginal smear records of the 15 grouped females to 3 groups of 5 females. The estrous cycles of the grouped females in both the first and second 30-day phases of the study were found to statistically differ from both random control groups.

In the second experiment of this study, 40 female rats (20 in the spring and 20 in the fall) were individually housed and separated physically from other female rats. Four subsets of five rats each shared a common source of recirculated air. Control groups were generated by "recasting the vaginal smear records of these same rats into a random set of four groups of five each" (p. 271). These rats were said to exhibit the number of regular four-day cycles comparable to those seen in grouped female rats. Since there was no physical or presumed acoustic contact among the rats, volatile cues, likely mediated through the olfactory system, were said to be involved (McClintock, 1981). As stated by the author, "When both the vaginal and lordosis components were measured, the estrous phases of the members of a group begin to coincide two or three cycles after the females were grouped together. That is, the majority of the group came into estrus on a particular day and then again repeated at 4-day intervals thereafter" (p. 246).

A theoretical mechanism based upon a coupled-oscillator model was subsequently evoked as the likely basis of such synchrony (McClintock, 1984; Schank and McClintock, 1992). According to this model, a phero-mone released during the follicular phase of the cycle would serve to shorten cycles, whereas one released near the time of ovulation would lengthen them. Early simulations demonstrated that such a mechanism was capable of synchronizing cycles within a group, although the antici-pated degree of synchrony was only slightly above that expected by chance (Schank and McClintock, 1992).

Since the publication of this work, the key assumptions of the model have been questioned. Schank (2000b), the primary author of the paper that described the application of the coupled-oscillator model for assess-ing estrous synchrony, points out the following:

> The simulation results presented in Schank and McClintock (1992) showed that estrous synchrony among female rats was a very weak phe-nomenon at best. A subsequent test of the coupled-oscillator hypothesis in rats failed to verify its assumptions (Schank and McClintock, 1997). A recent computer-simulation reanalysis of McClintock's (1978) report of synchrony among female rats showed that the level of synchrony reported could not be distinguished from chance (Schank, 2001d). Nor does syn-chrony appear to occur (Schank, 2000a) in the only other rodent species (golden hamsters, *Mesocricetus auratus*) in which it has been reported (Handelmann et al., 1980). Thus, while a coupled-oscillator mechanism could synchronize ovarian cycles, there is no evidence for this mechanism in mammals. (p. 844)

In a study that failed to replicate the estrous synchrony, Schank candidly states the following (Schank, 2001d):

> From the beginning, there has been no compelling theoretical reason to expect to find synchrony among female rats (McClintock, 1981), nor has the phenomenon ever been replicated. A test of the coupled-oscillator model of synchrony among Norway rats failed to confirm its basic assumptions (Schank et al., 1995). Moreover, that study found an opposite effect; airborne odors from the ovulatory phase of the estrous cycle reduced cycle lengths rather than lengthening them (Schank and McClintock, 1997). (p. 133)

He goes on to state, in relation to the original rat study, the following:

> A cursory examination of Figs. 1 and 2 in McClintock (1978) indicates that the main source of change in synchrony level was due to a decrease in synchrony in the "random" control groups with only a small increase in the synchrony of grouped (treatment) animals. Because the control groups were randomly constructed from the vaginal smear records of the experimental animals, the expectation is that on average they should not have changed in level of synchrony over time. Nevertheless, all four random control groups decreased in synchrony over time. This is a statistical artifact and may have been due to a small sample size (i.e., only three groups for each condition).
>
> In experiment II, although the level of synchrony was the same as in experiment I, we do not know whether the effect was due to synchronization of the cycles in the treatment condition or to the decrease in synchrony of the random controls. Again, a likely statistical artifact of using small sample sizes. (p. 134)

Schank (2001c) subsequently performed a comparatively large study to establish whether estrous synchrony was a viable phenomenon in rats. In this study he employed 88 rats (22 groups of 4 sisters) that were removed, at 12 days of age, to an experimental room and housed with their dams in specially designed habitats until they were 30 days of age. These habitats had an area to which only the dam could go. Weaning occurred at 30 days of age, and 4 female pups from each litter were separated from their littermates and housed at that time together in polypropylene maternity cages. Vaginal smears were collected beginning around the time of vaginal opening (i.e., 25 days of age) until the age of 70 days. In the special habitats, male urine or distilled water was presented in the section of the habitat only accessible by the dam, twice per day, for 16 successive days, that is, when the pups were 14 to 29 days of age. Since no influences of male urine were observed, the data from the urine and water exposure groups were combined. Schank found that estrous synchrony did not occur within the 22 groups of females of his study; if anything, there was a slight decrease in synchrony over time that probably reflected inter-female differences in mean cycle length. The Monte Carlo simulations also found that the degree of synchrony observed during the last estrus was consistent with that expected by chance and was quantitatively similar to the degree of synchrony observed at first estrus.

Such observations imply that estrous synchrony is a questionable entity in rats. Importantly, no attempts at isolating stimuli that might be considered a pheromone have been made. A similar state of affairs exists for reports of menstrual synchrony in nonhuman primates, as noted below, and in humans, as discussed in Chapter 7.

Estrous Synchrony in Nonhuman Primates

Estrous synchrony similar to that described above for rats has been reported for several nonhuman primates. Among the latter are the Hamadryas baboon, *Papio hamadryas* (Zinner et al., 1994), the chimpanzee, *Pan troglodytes schweinfurthii* (Wallis, 1985), and the golden lion tamarin, *Leontopithecus rosalia* (French and Stribley, 1985). However, as is the case with the aforementioned rat studies, as well as studies reporting menstrual synchrony in humans (Chapter 7), the statistical models used to identify synchrony actually bias data toward synchrony. When Schank (2001a) reanalyzed data extractable from two of these studies (Wallis, 1985; French and Stribley, 1985) using computer simulation models that overcome such biases, no evidence for synchrony was found. Similarly, using a data analysis procedure that was not statistically biased toward synchrony, Matsumoto-Oda et al. (2007) evaluated the timing and frequency of estrus, as measured by the turgidity of ano-genital swellings, of a long-studied group of chimpanzee females within the Mahale Mountains National Park, Tanzania. Data from May to January in nine different years from 1981 to 1994 were assessed. No statistical evidence of estrous synchrony was found, with estrus appearing randomly in five of the nine years evaluated. In the other four years, more *asynchrony* than expected by chance was noted, in accord with an earlier study of ring-tailed lemur (*Lemur catta*) that employed a Monte Carlo based statistical analysis (Pereira, 1991).

It is apparent from the material reviewed in this chapter that no unique single chemical that can be considered a reliable "priming" pheromone has been identified in mammals. While multiple chemicals have been isolated from the urine of male mice that can alter the time of puberty of females, their effects are likely interactive with one another and, as noted by Jemiolo et al. (1989), may depend upon ratios of multiple urinary components. The question of whether urinary chemicals placed on the oro-nasal groove of rodents are inducing stress-related responses is not clear, although there is no doubt that a number of behavioral and endocrine responses of mammals to chemical stimuli are mediated by stress-related pathways. Importantly, most phenomena attributed to pheromones are unlikely present in natural populations, in many cases likely being artifacts or epiphenomena (Labov, 1981; Bronson, 1979; Bronson and Coquelin, 1979; Rogers and Beauchamp, 1976).

SEVEN

Human Pheromones

Wonder not, Rufus, why none of the opposite sex
wishes to place her dainty thighs beneath you,
not even if you undermine her virtue with gifts of choice
silk or the enticement of a pellucid gem.
You are being hurt by an ugly rumour which asserts
that beneath your armpits dwells a ferocious goat.
This they fear, and no wonder; for it's a right rank
beast that no pretty girl will go to bed with.
So either get rid of this painful affront to the nostrils
or cease to wonder why the ladies flee.

Gaius Valerius Catullus (87–54 BC)

Whether humans have pheromones was listed by *Science* magazine as one of the top 100 outstanding scientific questions of 2005 (Anonymous, 2005). Over the course of the past three decades, numerous claims have been made that human pheromones influence sexual attraction, mate selection, moods, behaviors, seating patterns in offices, and endocrine function in both sexes. Although no chemical compounds have been isolated, in some cases specific androgen-related steroids have been assumed a priori as being human pheromones based upon their presence in human axillae and their purported pheromonal role in pigs. One group has claimed that humans possess a functional vomeronasal organ that responds to such agents in a sexually dimorphic manner (Monti-Bloch et al., 1994; Berliner et al., 1996).

As will be shown in this chapter, evidence for the existence of human pheromones is weak on empirical, conceptual, and methodological grounds. While odors and fragrances, like music and lighting, can alter mood states and physiological arousal, it is debatable whether agents purported to be pheromones *uniquely* alter such states. The same is true for putative pheromones said to alter endocrine function, as exemplified by questionable claims of their ability to synchronize menstrual cycles, a topic considered in great detail later in this chapter.

Sources of Putative Human Pheromones

Humans excrete or secrete a wide range of chemicals via their urine, breath, genitalia, saliva, and specialized skin glands. Following Alex Comfort's influential 1971 *Nature* paper entitled "Likelihood of Human Pheromones," investigators have largely confined their search for pheromones to secretions from the apocrine glands within the armpits. Most seem to have agreed with Grammer et al.'s 2005 statement, "The main producers of human pheromones are the apocrine glands located in the axillae and pubic region" (p. 136). This is in spite of the fact that—as noted in previous chapters—urine and sebaceous glands are the primary sources of putative pheromones in most nonhuman mammals.[1]

The apocrine glands are one of three major glands that excrete materials to the surface of the human skin (Figure 7.1). Among their secretions are lipids, including cholesterol, sterol esters, triglycerides, diglycerides, fatty acids, and wax esters. Apocrine secretions are largely odorless until acted upon by aerobic diphtheroid bacteria (Leyden et al., 1981). Despite being found in some of the same body regions as sebaceous glands, emptying into the shafts of hair follicles as part of the "apo-pilo-sebaceous unit," their highest density is within skin areas paralleling the evolutionary regression of terminal hair, such as the axillae and the perineum (Doty, 1981). These glands become functional around the time of puberty, releasing their secretions in relation to emotions associated with anxiety, fear, pain, or sexual arousal (Wilke et al., 2007). Some have suggested that their emulsifying properties serve an important thermoregulatory effect even in humans, although, in contrast to some other mammals (e.g., horses), their function is not thermoregulatory (Scott et al., 2001). Given their close functional association with anxiety and fear, it is conceivable

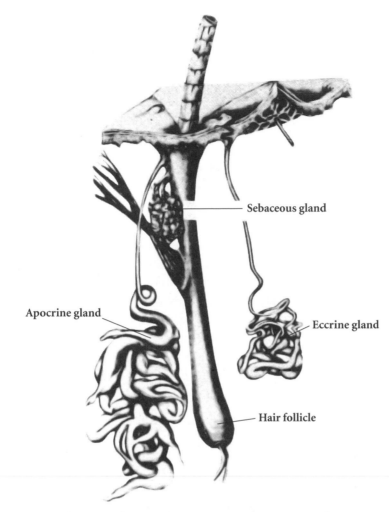

Sebaceous gland

Apocrine gland

Eccrine gland

Hair follicle

FIGURE 7.1. Schematic of a pilosebaceous unit with apocrine and eccrine sweat glands. Modified from Doty (1981).

that they represent a vestigial defense system useful at one time in warding off predators or unwanted conspecifics.

In contrast to the apocrine glands, the eccrine sweat glands play a significant role in regulating body temperature in humans. These three to four million glands are activated by the sympathetic nervous system during exercise and stress (Nicolaides, 1974) and are capable of secreting up to 3 l of aqueous solution per hour (Tobin, 2006). Their coiled tubules connect to the skin's surface independently of hair follicles. They are found in nearly all body regions, save the lip margins, nail beds, nipples, inner preputial surface, labia minora, glans penis, and glans clitoridis. Even though their secretions are generally inodorous, they can become odorous as a result of diet (e.g., garlic) and disease, including genetic disorders (e.g., fish odor syndrome).

The sebaceous glands secrete most of the lipids of the skin (e.g., triglycerides), as well as antimicrobial products (Nicolaides, 1974). Although there is room for debate, some believe that a major function of these glands is to keep the skin supple and waterproof. They are most dense on the forehead, face, and scalp, and are absent on the soles of the feet and palms of the hand. Specialized sebaceous glands are found in the eyelids, the ear canals, the nares, the lips, the buccal mucosae, the breasts, the prepuce, and the ano-genital region. In addition to producing sebum, these glands play an important role in the regulation of steroidogenesis, local androgen synthesis, skin barrier function, interaction with neuropeptides, and possibly the production of both anti- and pro-inflammatory compounds (Tobin, 2006). When not infected by bacteria, their secretions reportedly have a weak, pleasant odor. Like apocrine glands, these glands enlarge at the time of puberty, being influenced by gonadal hormones.

Potential Receptors for Putative Human Pheromones

The sensory means by which human pheromones might be detected are multiple. They include the olfactory, gustatory, trigeminal, and putative vomeronasal neural systems. Additionally, such agents could penetrate the brain directly via the olfactory mucosa or enter the general circulation by way of the vasculature of the nose, sinuses, oral cavity, and lungs (Doty, 2008). Although, as noted in Chapter 2, a number of molecular biologists are having second thoughts about the idea that mammalian pheromones only exert their effects via the vomeronasal organ (VNO), this organ is still viewed by most who believe in pheromones to be the main mediator of pheromonal activity. Hence, it is critical to understand the nature of the human VNO and whether it could be a sensory organ responsible for the transmission of putative human pheromones.

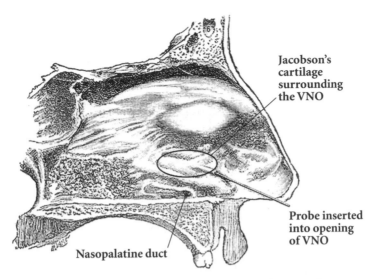

Jacobson's
cartilage
surrounding
the VNO

Probe inserted
into opening
of VNO

Nasopalatine duct

FIGURE 7.2.   Anatomical illustration by Potiquet (1891) showing the vomeronasal duct as well as the general location of the human vomeronasal organ (VNO)
into which it opens. Also shown is the nasopalatine duct, which connects to the
nasopalatine canal, a canal that in some animals opens into the VNO. In rare
instances this duct is found in humans, connecting the oral and nasal cavities.
Modified from Jacob et al. (2000).

The anatomy and physiology of the olfactory, gustatory, and trigeminal
systems are reviewed in detail elsewhere (e.g., Bartoshuk and Beauchamp,
1994; Doty, 2001, 2003a, b; Hummel and Welge-Lussen, 2006; Katz et
al., 2008; Rolls, 2005; Smith et al., 2008).

The human VNO is a pouch-like tube that opens into the nasal cavity
via a 0.2 to 2 mm diameter duct or pit located approximately 2 cm from
the naris where the septal cartilage joins the bony septum (Garcia-Velasco
and Mondragon, 1991; Stensaas et al., 1991; Moran et al., 1991; Meredith,
2001). This pouch, whose opening is depicted in Figure 7.2, is present in
the vast majority of adult humans; ranges in length from 2 to 33 mm; and
contains stratified, respiratory, and pseudostratified columnar epithelia
(Moran et al., 1995). The human VNO differs considerably from the
VNO of animals for which function has been demonstrated. For example,
it is sometimes found only on one side of the nose (Knecht et al., 2001)
and, in common with the VNO of adult Old World monkeys and apes,
lacks the full complement of cells found in functioning VNOs of other

mammals (Trotier et al., 2000; Bhatnagar et al., 2002; Smith et al., 2002). Few, if any, of its cells stain positively for olfactory marker protein, an index of functional olfactory and vomeronasal receptor cells. Although a small subset of bipolar cells stain positively for such neuronal markers as neuron-specific enolase (NSE), protein gene product (PGP 9.5), and soybean lectin (Trotier et al., 2000; Takami et al., 1993; Witt et al., 2002), proliferation antigens are not regularly expressed in the nuclei of cells located near the basement membrane. Nonetheless, positive reactions to CD44 suggest there is some cell differentiation and migration within sectors of the organ (Witt et al., 2002). The adult human VNO lacks the VNO-specific TRP2 membrane channel, a nonselective cation channel critical for VNO function in other mammals (Tirindelli et al., 1998), and most if not all human VNO receptor genes are pseudogenes presumed to have no function (Liman et al., 1999; Kouros-Mehr et al., 2001). The human VNO has no centrally projecting neural connection (i.e., no vomeronasal nerve) (Meisami et al., 1998; Trotier et al., 2000) and humans lack the central brain structure to which vomeronasal nerves ordinarily project (the accessory olfactory bulb) (Meisami and Bhatnagar, 1998). Comparative genetic studies suggest a functional human VNO was likely lost at the time of the divergence of New World monkeys at the end of the Eocene period (Evans, 2003).[2]

Taking all of this evidence together, the human VNO is perhaps more reasonably described as a VNO remnant. Nonetheless, a research team at the University of Utah reported, in a series of provocative studies, that humans possess a viable VNO responsive to "vomeropherins" (Monti-Bloch and Grosser, 1991; Berliner et al., 1996; Monti-Bloch et al., 1998; Grosser et al., 2000). This team, funded by a company that makes personal care products containing such agents, reported that several steroids, most notably androstadienone, produced local electrical potentials within the lumen of the human VNO. Such potentials were seemingly analogous to the odor-induced electro-olfactogram found in the main olfactory epithelium (Monti-Bloch and Grosser, 1991). They were (1) not induced by clove oil and un-named "conventional" odors; (2) not accompanied by conscious perception; (3) not obtained from surrounding respiratory epithelium; and (4) found to produce, in some cases, sexually dimorphic responses. Thus, androstadienone purportedly activated the female, but not the male, VNO, whereas estratetraenol had the opposite effect. The authors indicated that the compounds that induced the VNO activity produced only weak responses in the main olfactory epithelium. The VNO stimulation purportedly influenced, in sex-dependent ways, hormone secretion, the electroencephalograph (EEG), various autonomic nervous system measures, and psychological mood states (Monti-Bloch and Grosser, 1991; Stensaas et al., 1991; Moran et al., 1995; Monti-Bloch et al., 1994). For example, 35 minutes after the brief administration of androstadienone, decreases in respiratory and cardiac frequency, lowering of the galvanic skin response, an increase in alpha cortical activity, and an increase in body temperature were reported (Grosser et al., 2000). These investigators concluded "that this substance [androstadienone] can now be considered as a pheromone is indicated by its ability to alter behavior, as well as its gender specificity" (p. 295).

While provocative, this work has never been replicated and, as described later in this chapter, the clear-cut sexually dimorphic effects found for androstadienone and estratetraenol have not been seen in behavioral studies in which these agents have been inhaled. It is difficult to imagine that the responses observed by the Utah investigators were mediated by a functioning VNO system, given the sparseness of the bipolar cells that express neuro-specific enolase (NSE), which would be the most likely candidates for sensory receptor cells within this system, and the fact that humans lack a vomeronasal nerve and accessory olfactory bulb. In contrast to their claim of no androstadienone-related electrophysiological responses within the respiratory epithelium, others have subsequently found such responses at concentrations below 2 ppm (Boyle et al., 2006). This suggests the purported VNO effects could have been mediated via the trigeminal nerve. Activation of this nerve is well known to produce alterations in a range of autonomic nervous system parameters, including heart rate, respiration rate, blood pressure, nasal engorgement, mucus secretion, and epinephrine secretion (Doty and Cometto-Muñiz, 2003). A review of the factors that might explain the chemically induced electrophysiological responses observed within the human VNO is provided by Meredith (2001).

Whatever the basis for the responses observed by these authors following chemical or electrical stimulation of the human vomeronasal sac, this line of research has energized a number of investigators to focus attention on androstadienone as a potential human pheromone, since this agent was the most salient "vomeropherin" they found.

Studies of Human Axillary Secretions
....

Can Humans Discern Sex on the Basis of Axillary Odor?

Under the widespread assumption that pheromones exist and come from axillary secretions, it is logical to ask the question whether information critical for reproductive behavior can be sensed from such secretions. In a pioneering study, Russell (1976) had 13 women and 16 men wear T-shirts for 24 hours without bathing or using deodorants. The armpit regions of the T-shirts were then presented in a test session where (1) the subject's own T-shirt, (2) a male stranger's T-shirt, and (3) a female stranger's T-shirt served as stimuli. Each subject was asked to identify his or her own odor and then to report which of the two remaining odors came from a male. Nine of the 13 women and 13 of the 16 men performed both of these tasks correctly, suggesting to Russell that "at least the rudimentary communications of sexual discrimination and individual identification can be made on the basis of olfactory cues" (p. 521).

Since women have smaller apocrine glands than men (apocrine gland size correlates with body odor intensity) and commonly shave their axillae (reducing the area for bacterial activity and stimulus diffusion), the aforementioned judgments of male and female odors could be based on stimulus intensity or pleasantness, rather than on sex-specific factors present in the secretions. In other words, subjects may use the strategy of assigning stronger or more unpleasant odors to the male category and weaker or less unpleasant odors to the female category. This would be analogous to assigning, from a list of body weights, heavier weights to the male category and lighter weights to the female category.

To test this possibility, a series of studies was performed in the late 1970s in which axillary secretions were collected from gauze pads taped in the armpits of male and female donors for approximately 18 hours (Doty et al., 1978). These pads were then presented to 10 male and 10 female subjects, along with blank gauze pads, in 100 ml glass bottles, capped until sampling, with the following instructions:

You will be presented with a series of sniff bottles containing human sweat. We wish you to tell us [by smell] which sex each of the odor samples comes from. The set of odors may include samples from both men and women, or from only men or from only women. Thus, some may be from females, some from males, or, alternatively, all may be from females or all from males. Therefore, don't allow yourself to assume that some predetermined number of one or the other sex is represented.

In addition to being asked to identify the sex of the donors, the subjects were asked to estimate the relative intensity and pleasantness of the stimuli using a magnitude estimation procedure (Doty and Laing, 2003). In some cases, half of the stimuli were from men and half from women; in other cases, all were from women or all were from men. The results of the study in which axillary odors from both men and women were presented are shown in Figure 7.3. Note that four of the five male odors were correctly identified as male by at least half of the subjects, whereas three of the four female odors were correctly identified as female by at least half of the subjects, and that intensity was strongly related to these assignments. The blank (B) stimulus was typically assigned to the female category.

The results of the study in which only female odors were presented are shown in Figure 7.4. It is clear from these data that even when all odors came from women, the strongest ones were assigned to the male category and the weakest ones to the female category, in accord with the hypothesis that intensity was the primary basis for the sex classifications. In this case, all subjects assigned the blank to the female category.

If axillary odors serve as sex attractants without extensive conditioning, one would expect that male axillary odors would be rated relatively more pleasant by women than by men, and that the reverse would be true for female axillary odors. Such relations were not observed. Thus, both sexes rated male odors as more intense and less pleasant than female odors, with the relative magnitude of the responses of the two sexes being similar (rs > 0.90) (Doty et al., 1978). The overall intensity and pleasantness estimates were strongly and inversely related to one another ($r = -0.94$, $p < 0.001$). Such general associations are not culture-specific (Schleidt et al., 1981). Moreover, no differences in pleasantness or intensity are perceived when equivalent quantities of male and female apocrine

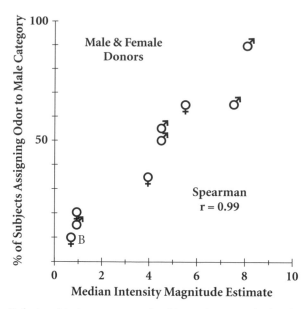

FIGURE 7.3. Relationship between perceived intensity, sex of odor donor, and percent of subjects assigning each axillary odor to the male gender category. Male and female symbols signify the sex of the odor donor. B indicates blank control. Data from Doty et al. (1978).

sweat are incubated in vitro with aerobic coryneform bacteria, which produce the characteristic underarm odor (Gower et al., 1994).

A finding that axillary hygiene markedly decreases the ability of subjects to correctly assign gender to axillary odors is in accord with the hypothesis that such ability is based upon the relative intensity or pleasantness of the stimuli (Schleidt, 1980). When equated for volume, axillary odors are not differentially pleasant to males and females, as might be expected if they contained sexually dimorphic pheromones associated with attraction or repulsion (Zeng et al., 1996). It is important to note that *no sex differences are present in the odorless precursor proteins within apocrine glands ultimately responsible for axillary odor,* although males may be more prone than females to release their contents in relation to stressful stimuli (Spielman et al., 1998). The extent to which learning influences hedonic responses to axillary odors is unknown, although children who can identify their source as the armpits rate such odors as more intense and less pleasant than children who cannot identify their

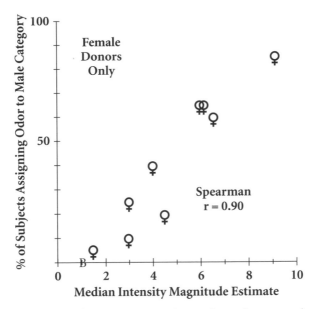

FIGURE 7.4. Relationship between perceived intensity and percent of subjects assigning each axillary odor to the male gender category. Note, as indicated by the female symbols, that all of the axillary odors came from women. B indicates a blank control. Data from Doty et al. (1978).

source, implying some role of learning. In accord with adult findings, adolescent girls rate axillary odors as more intense and less pleasant than do adolescent boys (Stevenson and Repacholi, 2003).

It should be stressed that axillary odors are significantly influenced by diet. In one study 17 men were placed on a meat diet for two weeks and on a non-meat diet for two weeks, with the order of the diets counterbalanced in time (Havlicek and Lenochova, 2006). Axillary secretions were collected using gauze pads worn for 24 hours at the end of the dietary periods. The odors of donors on the meat diet were rated as less attractive, less pleasant, and more intense than the odors from the donors on the non-meat diet, demonstrating the important role of diet in altering both the intensity and pleasantness of axillary odors. Such observations suggest the possibility that some sex differences observed in axillary odors are confounded by sex differences in dietary habits. Recent population studies clearly indicate that men eat much more meat than women (Shiferaw et al., 2008).

Do Human Axillary Secretions Influence the
Sexual Attractiveness of Others?

Intense axillary odors tend to repel others, as illustrated by Catullus's lyric at the beginning of the chapter. The question as to whether weak axillary odors, perhaps even sub-threshold ones, influence sexual attractiveness has received only cursory study.

Thorne et al. (2002) sought "to determine whether naturally occurring pheromones might act as sexual attractants in humans" (p. 292). To test this hypothesis, 32 college women, half of whom were taking oral contraceptives, rated male vignette characters and photographs of male faces for their attractiveness in the presence or absence of segments of gauze pads that had been worn overnight by men. The stimulus and control (blank gauze pads) stimuli were hidden out of sight in separate test booths where the rating tasks were performed. Each subject was successively tested in the stimulus and control test booths during menses and 14 to 21 days later, resulting in 4 repeated test conditions counterbalanced in order. The subjects were told that the purpose of the study was to determine the effects of blood glucose on impression formation at different stages of the menstrual cycle and were given a sugarless grape beverage to drink before each test.

Under the axillary secretion exposure condition, the hypothetical men were rated as better looking and more sexually attractive only by the women taking oral contraceptives. Under the control exposure condition, the oral contraceptive group actually rated the men as less better looking and less sexually attractive than did those not taking such contraceptives. The men were rated as more self-assured and having a nicer body when the pads containing axillary secretions were present than when the control pads were present, regardless of cycle phase or contraceptive group.

The results of this study, which has never been replicated, are difficult to interpret for a number of reasons. First, as noted by Grammer et al. (2005), the study "simply required participants to rate the attractiveness of hypothetical opposite-sex characters based on written descriptions and photographs. The ecological validity of such laboratory-based studies is questionable" (p. 138). Second, many ratings were made on a number of dimensions, leading to the likelihood that some of the observed effects were due to chance. Third, it is not clear why the hypothetical men should be rated as better looking and more sexually attractive only by the women taking oral contraceptives, although mood can be influenced by oral contraceptives (e.g., during menses, oral contraceptive users are typically less depressed and have more stable moods) (Oinonen and Mazmanian, 2002). Fourth, while a statistical comparison was made with the no-odor control, the question remains as to whether the observed effects are specific to axillary secretions. Would other stimuli produce the same effect? It is of interest that the authors assumed that wearing the gauze pads in the armpits for 8 to 12 hours did not result in the production of a discernable odor, as the women "were unknowingly exposed to pure male axillary secretions (unaffected by coryneform bacteria and therefore not consciously odorous)" (p. 292).

As described later in this chapter, most research seeking to determine the effects of human pheromones on psychological measures has not employed natural secretions sampled from the axillae. Instead, single commercially available agents, such as androstenone, have been used that are assumed a priori to be human pheromones.

Do Human Axillary Secretions Influence Mood?

A number of stimuli can influence human moods, emotions, and feelings. The question is whether human secretions or excretions can do so in a manner that can be construed as "pheromonal." What is the evidence that volatile or nonvolatile agents of human origin, most notably those from the axillae, alter emotion and mood in ways different from that of non-pheromonal odorants or other sensory stimuli?

In a study that received considerable attention in the popular press, Chen and Haviland-Jones (1999) reported that axillary odors influence moods in ways that depend upon the age and sex of the odor donors. Gauze pads were worn in the axillae for several days by 5 prepubertal girls, 5 prepubertal boys, 5 college women, 5 college men, 5 older women, and 5 older men. Ratings of mood were made before and after volunteers briefly smelled the odor-laden pads placed in Petri dishes. More than 300 volunteers took part, with a given subject smelling only one type of stimulus in the mood assessment element of the study. The researchers concluded that "exposure to underarm odors of older women, women, and older adults, led to a greater reduction in depressive mood than exposure to underarm odors of young men, men, and young adults." According to Black (2001), Chen was quoted in the popular press as stating "old women had an uplifting effect" on the moods of others, whereas Haviland-Jones was said to have speculated that "eau de grandmere" functions to signal a happy mood state (p. 248).

Unfortunately, this study is fraught with problems. Black (2001) points out, after carefully examining their study, that the average positive mood actually declined in subjects exposed to the axillary odors from each of the six donor groups, contrary to the aforementioned comments of the authors, and that employment of the Differential Emotion Scale (Izard et al., 1993) as a measure of mood was inappropriate. According to Black, this scale is not designed to assess short-term changes in mood, but how often different moods are experienced, suggesting the subjects were likely confused as to how to respond on the post-odor test administration. Most importantly, Black noted that while Chen and Haviland-Jones collected data from a control condition (pads just left in homes but not placed in the armpit), they did not appropriately use these data in their analyses. When the mean data of each of the target groups were compared with those of the controls, no statistically significant effects emerged, leading Black to conclude that "it is clear that their work provides no evidence for their conclusion that certain human odors can decrease depressive mood" (p. 216).

Androstenone, Androstenol, Androstadienone, Copulin, and Related Agents as Human Pheromones

Of all substances purported a priori to be human pheromones, the androstenones are the most universal, spawning dozens of studies on this topic. Those traditionally most widely touted are 5α-androst-16-en-3α-one (androstenone), 5α-androst-16-en-3α-ol (androstenol), and androsta-4, 16-dien-3-one (androstadienone). These are putative pig pheromones described in Chapter 5. Their chemical structures and synthetic pathways are shown in Figure 7.5.

The historical reasons for assuming such agents are human pheromones are multiple. First, they are among the few identified compounds that have been directly linked to mammalian reproductive behavior (i.e.,

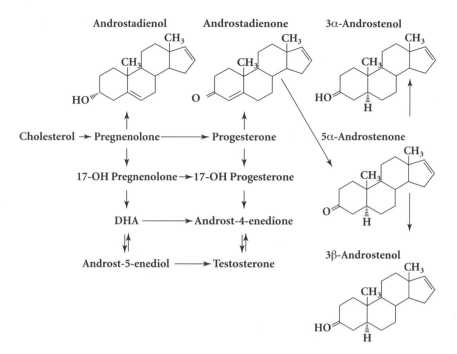

Androstadienol Androstadienone 3α-Androstenol

5α-Androstenone

3β-Androstenol

Cholesterol → Pregnenolone ──────→ Progesterone

17-OH Pregnenolone → 17-OH Progesterone

DHA ──────→ Androst-4-enedione

Androst-5-enediol ──────→ Testosterone

FIGURE 7.5. Biosynthetic pathways for androst-16-enes and androgens in the boar testis. Modified from Kwan et al. (1987).

lordosis in the female pig). Second, being steroids they fit the pheromone concept of an "externally secreted hormone." Third, they are reportedly found, albeit at low levels, in human urine, axillary apocrine sweat, saliva, and semen (Kwan et al., 1992; Brooksbank and Haslewood, 1961; Gustavson et al., 1987; Brooksbank et al., 1974; Nixon et al., 1988), making them potentially available to the external milieu for transfer from one person to another. Fourth, they typically have musk-like or urine-like smells to those who can smell them, reinforcing the notion of their animal-like nature and the folklore that musks are social attractants in humans (Gower et al., 1985; Le Magnen, 1952b; Kloek, 1961). Fifth, they occur in higher concentrations in men than in women, implying sexual dimorphism (Gower and Ruparelia, 1993; Lundström and Olsson, 2005). Sixth, women are more sensitive, on average, to these agents than men (Doty, 1986a), in accord with the notion of pheromonal sexual dimorphism. Seventh, in light of studies suggesting that humans can distinguish between the sexes to some degree on the basis of axillary and breath odors (Doty et al., 1978, 1982a; Hold and Schleidt, 1977), it is conceivable that these steroids serve to make this possible. Finally, in the absence of any known function, a pheromonal function would seem to fill the void. As noted by Gustavson et al. (1987), "Because androstenol has no known function in humans, these findings have suggested to several investigators that the steroid may function as a human pheromone" (p. 210).

While such arguments seem logical, closer scrutiny reveals problems. First, none of these steroids contributes much to the generation of prototypical body odor, which arises largely from a mixture of C_6–C_{11} normal, branched, and unsaturated acids (Zeng et al., 1991, 1992; Hasegawa et al., 2004). Second, the amount of such steroids in the axillae is low and highly variable. Using capillary gas chromatography/mass spectrometry with specific ion monitoring, Nixon et al. (1988) found that only 10 of 24 men had androstenone in their axillary hair and that no relationship was present between the age of the donors and presence of the steroid. Others, using different analytical methods, have reported finding no androstenone in samples of fresh apocrine sweat or secretions sampled by sterile gauze pads (Bird and Gower, 1981; Labows et al., 1979). Before bacterial action, androstadienone (AND) levels are too low to be detected in the axillae by smell (Labows, 1988; Gower et al., 1994). Third, it does not follow that simply because these compounds are found in body fluids or axillae that they communicate meaningful social information or influence reproductive processes in humans. Indeed, androstenone and androstenol are somewhat ubiquitous in the animal and plant kingdoms, being found even in the roots of vegetables such as parsnip and celery (Claus and Hoppen, 1979). In one study, for example, androsterone was found in 60 to 80% of the plant species investigated (Janeczko and Skoczowski, 2005). Fourth, it is not clear why having musky or urine-like smells should qualify such compounds as human pheromones. The fact that a significant number of persons cannot smell these chemicals (Amoore et al., 1977; Ohloff et al., 1983; Koelega, 1980), along with evidence most who do so find them repulsive or unpleasant (Gower et al., 1985; Jacob et al., 2006; Koelega, 1980), would seem to limit their value in social interactions that would be considered as reflecting "sex pheromones." In the case of androstadienone, one report suggests that repeated exposure results in an increase in its perceived unpleasantness (Boulkroune et al., 2007). Fifth, sex differences and subtle menstrual cycle-related fluctuations are present for a wide range of odorants, including synthetic ones, so there seems to be nothing special about these agents in this regard (Doty, 1986a; Doty and Cameron, 2009; Doty et al., 1981, 1982b). Sixth, as described in the previous section of this chapter, the ability of humans to determine the sex of another human on the basis of axillary odors, as well as other odors common to the sexes such as breath odors, appears to be dependent upon the intensity of the involved odors, not on intrinsic qualities of the secretions (Doty et al., 1978, 1982a). Finally, it does not follow that a lack of known function increases the likelihood an agent is a human pheromone.

If pheromones are species-specific, which is inherent in nearly all pheromone definitions (Chapter 2), then assuming these steroids are the same in humans as in pigs is an oxymoron. Are women, in fact, attracted to the odors of male pigs or more willing to have sex in the presence of such odors? Are birth rates or other indices of sexual behavior higher in states or counties with pig farms? Even if one drops species specificity from the pheromone definition, if the behaviors attributed to the agents differ markedly between species, it would seem unlikely that they have a homologous relationship.

Key studies reporting the influences of these agents on human behavioral and physiological responses are described in the following pages. Although numerous claims have been made that these steroids uniquely alter human mood and behavior, including the feelings people have about one another, their uniqueness is debatable, control odors are rarely employed, and results are contradictory. Some have concluded that the effects of these steroids are actually nonconscious, thereby negating concerns that a number of persons cannot smell them. While high supraphysiological levels of androstenone or related steroids may have pharmacological effects via their absorption into the bloodstream, it is debatable whether such absorption would occur at concentrations encountered in everyday life.

Influences of Androstenone on Seating Choices

In a widely cited study, Kirk-Smith and Booth (1980) reported that women entering a 12-chair dental office waiting room were more likely to sit in a chair odorized by androstenone than in chairs not so odorized, thereby exhibiting attraction to the stimulus. Men, on the other hand, were

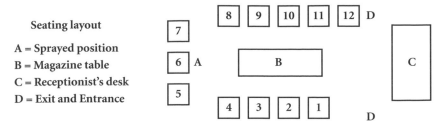

FIGURE 7.6.   Office seating arrangement of chairs in study by Kirk-Smith and Booth (1980) in which androstenone was said to attract women and repel men.

found to be less likely to sit in such a chair when it was sprayed with one of the three concentrations of androstenone that was employed (32 µg). The layout of the chairs in the small office waiting room is shown in Figure 7.6. The target chair (#6), which was the farthest one from the reception desk, was chosen for stimulus application because under 4 days of initial observation trials it was never sat in by women, although it was on occasion sat in by men. The results of this study, which in total was comprised of 21 days of observation in addition to the initial 4 days of control trials, are presented in Table 7.1.

As shown in this table, the frequencies of the use of the seats are relatively low, regardless of the odor condition. A 3 (seat type: sprayed, adjacent, other) $\times$ 4 (treatment condition: 0, 3.2 µg, 16 µg, 32 µg) χ^2 test of association performed for each sex separately was not statistically significant for either men (p = 0.10) or women (p = 0.07). However, significant differences were present between the controls and both the 3.2 µg and 32 µg odor conditions for the women (respective ps < 0.05 and 0.01), and between the controls and the 32 µg odor condition for the men (p < 0.05). The reason why such effects were not observed for the 16 µg odor condition is unclear, although the authors suggest this can be attributed to more crowding in the waiting room on the 16 µg test days.

Numerous methodological problems cloud this study. First, no control odors beyond a blank were evaluated, so there is no way of knowing whether the reported effect, if in fact true, is unique to androstenone. Second, an unknown number of repeat patients were recorded during the study period, violating the assumption of sample independence required by χ^2 analysis. Third, a regression-toward-the-mean effect may have biased the findings, given that the target seat was chosen on the basis of a

TABLE 7.1.   *Seating Distribution of Patients in a Dental Waiting Room Relative to Target (Odorized Seat, #6)*

	Control		3.2 μg		16 μg		32 μg		A-C DIFF
	No.	%	No.	%	No.	%	No.	%	(%)
Men									
Sprayed seat	20	11.05	6	6.82	16	11.19	5	8.20	−2.31
Two adjacent seats	29	16.02	12	13.64	13	9.09	2	3.28	−7.35
Other seats	132	72.93	70	79.54	114	79.72	54	88.52	+9.66
Totals	181		88		143		61		
Women									
Sprayed seat	6	4.17	6	9.23	7	7.95	8	11.27	+7.11
Two adjacent seats	4	2.78	7	10.77	5	5.68	7	9.86	+7.99
Other seats	134	90.05	52	80.00	76	86.36	56	78.87	−11.18
Totals	144		65		88		71		

Source: From Kirk-Smith and Booth (1980).

short sampling period during which it was not occupied by women. Fourth, the seat chosen for assessment is distal to all other seats and likely would be disproportionately chosen during periods when other seats were occupied, further eroding the independence of the measure. Fifth, the control days were clustered at the end and beginning of the other test days, potentially introducing temporal confounds. Sixth, in some cases the chair was sprayed with androstenone on the evening before the clinic and in other cases on the morning before the clinic, which presumably would affect the intensity of the stimulus during the test day. Seventh, since the stimuli were used for two days after application, one wonders whether odorant concentration changed over time or was influenced by prior persons sitting in the chair or by persons changing positions. Eighth, it is not clear why the stimulus would not diffuse generally into the room, thereby reducing its ability to be localized to a specific seat. Ninth, the number of seats occupied by children or by the elderly is not indicated, both of which would likely influence seating patterns. Finally, if there is ecological validity to the notion that axillary steroids influence human behavior, then one would expect interactions between the steroids on the persons within the waiting room and the androstenone on the seats. It would seem that a more appropriate experimental design would be to have odorized the experimental and control seats randomly and to have then determined the seating distribution relative to the odorized seats.

In contrast to the aforementioned findings, Gustavson et al. (1987) found no evidence that androstenone influenced the choice of a restroom stall by either men or women in male and female restrooms on different floors of the same dormitory. In their five-week study, the first, third, and fifth weeks served as control baselines. Androstenone was present during the fourth week and its related alcohol, androstenol, during the second week. Ironically, androstenone was considered a control odor for androstenol, as the authors assumed that it is not contained in human axillary sweat and therefore would make a sound control for androstenol (Gower and Ruparelia, 1993; Brooksbank et al., 1974). The sample of this study consisted of 480 uses of the stalls, implying that some subjects were recorded more than once, again violating the assumption of independent samples required for the χ^2 test. Stalls odorized with androstenol were found to be used less often by men, but not by women.

More recently, Pause (2004) instructed 40 homosexual men and 39 heterosexual women to sit in an "experimental" waiting room in their laboratory in which 4 chairs were positioned. This study overcame a number of the problems of the aforementioned studies, such as potential repeated visits by the same subjects to the test situation. Prior to entering the room, the subjects were in an adjacent room, where half were asked to "arouse themselves sexually by making use of their own fantasies" (p. 26). The other half "had to try to give themselves an appetite for food" (p. 26). In both cases, heart rate was monitored and found to be elevated after the instructions, an increase that the authors interpreted as affirming their effectiveness. Ten minutes after a given subject reported having reached "the relevant motivational state," he or she was asked to go into the experimental waiting room, whose furniture arrangement is schematically depicted in Figure 7.7. Androstenone, 78 μg diluted in 1 ml propanediol, had been "dripped on an absorbent piece of paper which was stuck to the back of the outer left chair <C 1>" (p. 26). The androstenone was renewed daily, and at the end of the test session the androstenone threshold was determined for each subject. The authors surmised that because the concentration of androstenone was near the detection threshold level, only subjects very sensitive to androstenone would be likely to have noticed its odor.

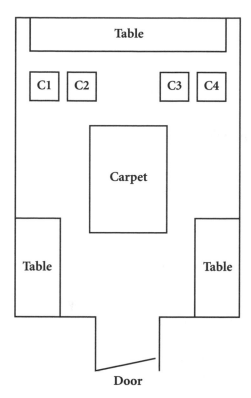

FIGURE 7.7.   Office seating arrangement of chairs in study by Pause (2004) in which androstenone was said to attract women. The stimulus was placed on the back of the C1 chair. Modified from Pause (2004).

For analysis, the data were compared to data obtained from a pre-study in which 51 women and 41 men had been asked to take a seat in the waiting room until a brief olfactory test was prepared. In this pre-study, 13.6% chose one or the other chairs on the left (C 1 and C 2). In the study proper, 27.8% chose chairs on the left (i.e., on the odorized side), a difference significant from the 13.6% value of the pre-study group. Unlike the Kirk-Smith study, no sex differences were noted and the sexual orientation of the subjects did not appear to influence the test results. However, the sexual orientation of the subjects in the pre-study was not mentioned. A correlation was noted between lower androstenone olfactory thresholds and the tendency to choose the seats on the left side of the waiting room.

Pause is circumspect about the findings of her study and acknowledges some of its limitations. For example, she notes that, in common with earlier studies, no control odors were employed and therefore it is not known whether the effects would occur for agents other than androstenone. Assuming that the presumptive pheromone works at an unconscious level, she points out that the tendency of those with lower androstenone thresholds to choose the left side of the room may implicate involvement of the odor of androstenone. She also points out the paradox as to why women seem to approach androstenone yet, in another study, judge themselves and men as less sexy in its presence (Filsinger et al., 1984). Other concerns regarding this work include the fact that waiting 10 minutes after a subject achieved either putative sexual or hunger arousal to evaluate the seating preferences may mitigate any influences of such states on potential responses to androstenone, and that the controls apparently were not asked to similarly motivate themselves before entering the room. One might question why these manipulations would influence responses to androstenone in the first place. Like the previous studies, no randomization of the presentation of the stimuli was made across the various seats and it is not clear why the stimulus would not diffuse over some range of space. Importantly, the composition of the experimental group differed from the pre-study group to which it was compared in several ways, presumably including more homosexuals. Although no differences in seating preferences were noted between the homosexual men and the heterosexual women, it is known that disproportionately more left-handers are found in homosexual populations (Lalumiere et al., 2000). Left-handers, in turn, are somewhat more likely than right-handers to choose left-positioned seats over right-positioned seats in some settings (e.g., auditoriums) (Karev, 2000).

Influences of Androstenol, Androstenone, and Copulin on the Evaluation of Others

In a widely cited study, Cowley et al. (1977) had 183 students of an introductory psychology class rate a set of six hypothetical applicants for a job, described in detail in a written narrative, on a series of traits related to job performance. Paper surgical masks were worn during the task that were either untreated, impregnated with androstenol, or impregnated with copulin, the concoction of fatty acids from rhesus monkey vaginal secretions which, as noted in Chapter 5, is a snark even when monkeys are concerned (Bonsall and Michael, 1971; Keverne and Michael, 1971). The traits of the hypothetical applicants, rated on a 5-point scale, were divided into favorable (e.g., "is dependent in what he says/does," "has good organizing ability") and unfavorable (e.g., "interests too narrow for the job," "lacks stability") categories. The six applicants varied on multiple personality and experience dimensions, with half being men and half being women.

Neither the androsterone nor the copulin treatments influenced the ratings of the unfavorable traits given to the hypothetical applicants. However, both of these substances reportedly influenced the ratings given to the positive traits, although no main effect of treatment was observed. The authors discuss a *non-significant* treatment by sex interaction that they state "arises mainly from the differences between the males and female students assessing candidate 'F'" (p. 164). The respective mean male and female ratings, reflecting a composite sum of the ratings across 14 favorable attributes, estimated from their Figure 1 and averaged across the 6 candidates for the 3 stimulus conditions were as follows: control— 35.7 and 37.2; androstenone—36.6 and 37.9; copulin—36.0 and 37.3. In all cases, women gave larger overall ratings than men but the relative differences were essentially the same for the 3 treatment conditions, ranging from 1.3 to 1.5 points. The error bars of their figure (which are not labeled but presumably reflect SEMs) extend more than ± 1 point in all cases.

The authors reported a significant treatment by candidate interaction (p < 0.05), noting that "in all instances where there is a Treatment × Candidate interaction, it is the female students who, in assessing male candidates, show wide variation in their judgments." They further note, "Equally striking is the reversal of the order of the scores when male students are assessing the male as opposed to female candidates. The males in the androstenol and fatty acid samples gave higher scores to the male candidates ('A', 'B', and 'C') and lower scores to the female candidates. The reverse is true for the control sample though neither of the differences is statistically significant" (p. 165).

One must question whether meaningful differences, in fact, exist across candidates for the three treatments. Presented in Table 7.2 are the estimated mean summed positive responses, averaged across sexes, for each treatment condition from Figure 1 of Cowley et al.'s paper. It is ap-

TABLE 7.2. *Estimated Mean Summed Positive Responses, Averaged across Sexes, for Cowley's Treatment Conditions*

Candidate	Control	Androstenol	Copulin	Androstenol, % change from control	Copulin, % change from control
A	37.7	38.5	36.3	+2.1	−3.7
B	32.5	34.5	35.1	+6.2	+8.0
C	38.4	40.4	38.5	+5.2	+0.3
D	41.1	40.3	40.6	−1.9	−1.2
E	34.2	31.9	32.7	−6.7	−4.4
F	37.2	36.6	36.4	−1.6	−2.2
Overall mean value	36.9	37.0	36.6	+0.3	−0.8

Note: Mean sum of positive attribute ratings of hypothetical job candidates given by subjects wearing paper surgical masks with no odor (untreated), androstenol, or copulin fatty acid rhesus monkey vaginal secretion mixture.

parent from this table that the mean ratings given to the hypothetical candidates differ little *across the treatment conditions* when all subjects are considered, with ratings under the androstenol and copulin conditions differing, on average, less than 1% from those of the control condition. Spearman correlation coefficients computed among the three treatment conditions suggest the ratings of the applicants were similar across candidates, regardless of the type of odorant embedded in the mask (all three rs = 0.89, p < 0.01).

Aside from statistical concerns, this study suffers from other problems. It is not clear whether the subjects were blinded to the purpose of the study. The task was quite artificial, bearing little resemblance to situations where a pheromone, if it were to exist, might be expected to work in social settings. If the effect were being mediated by the olfactory system, one might expect the odor's influences to decrease over time as a result of adaptation or habituation to the stimulus. Importantly, there is no strong theoretical basis to expect androstenone or copulin to influence the judgment of others, since the stimuli are not emanating from the target. This is attested to by the fact that the experimenters neither posed nor tested any hypotheses in this study. Even if the findings discussed by the authors were valid, they are complex, difficult to conceptualize, and seemingly dependent upon non-quantifiable and interacting traits of the hypothetical persons who were being evaluated.

In a similar but smaller study, Kirk-Smith et al. (1978) had 12 men and 12 women rate 16 photographs of normally clothed people, 4 photographs of animals, and 4 photographs of buildings on 15 9-point scales derived from Osgood's semantic differential (Osgood et al., 1957) under 2 test conditions. In one, a surgical mask was impregnated with 0.3 mg of androstenol. In the other, no odor was added to the mask. The order of the sessions was counterbalanced across subjects. The authors reported that under the androstenol odor condition both men and women rated the photographed women, but not the men, as sexier, more attractive, and better than under the no odor condition. The assumption was made that the effect was via the olfactory system, since the authors excluded from analysis a subject who, after the experiment, was found unable to smell the androstenol. A second subject was omitted from analysis because of experimenter error. The authors concluded, "This experiment is the first clear characterization of an effect of naturally secreted pure odour on measurements of sexual and agonistic relations among human subjects" (p. 379).

The Kirk-Smith study shares most of the same limitations of the Cowley et al. study, namely, an artificial test situation, lack of a theoretical basis, constant presentation of the stimulus, and failure to test any specific hypothesis. It is not clear why, for example, both sexes would rate the women as sexier and more attractive if androstenol was an agent involved in heterosexual attraction or repulsion. Unlike the Cowley et al. study, in which copulin was also assessed, only one type of odorant was employed, making it difficult to ascertain whether the observed effects, if indeed valid, would be present for any other type of stimulus. As pointed out by Gower et al. (1988), musks having smells similar to those of androstenone and androstenol could be used as control stimuli to assess the specificity of such agents.

To address a number of these problems, Black and Biron (1982) examined the influences of androstenol in a situation where direct interaction occurred between pairs of the opposite sex. One member of each pair was a confederate of the experimenter. On some occasions the confederate was odorized with androstenol, on other occasions with the odor of a synthetic musk, and on still other occasions with no odor at all. Thus, more than one musk-like agent was used, allowing for a test of the specificity of androstenol. The confederate was unaware of which odorant was applied on a given test day. The 78 subjects were randomized to the treatment conditions and placed next to the opposite-sex confederate during a 15-minute session in which they were shown slides of animals and flowers. After the slide presentation, the pair members were separated and asked to fill out a questionnaire whose ostensible purpose was to rate the pleasantness of the slides. Also included were questions concerning the attractiveness of the other pair member (confederate) on a 10-point scale. Additionally, the subjects were tested for their ability to smell the two study odorants. While all subjects could detect the synthetic musk, data from 7 who were unable to smell androstenol were excluded from analysis.

Even though the male confederate was judged as being significantly more attractive to the females than was the female confederate to the males, neither the androstenol nor the synthetic musk influenced the subjects' ratings. The authors concluded, "The present study is the first test of the pheromonal properties of androstenol in a naturalistic situation. The results obtained fail to support the hypothesis that androstenol can influence the judgments of physical attractiveness compared with a control condition" (p. 329). They further noted, "The failure to find positive effects under these circumstances suggests that it is premature to classify androstenol as a human pheromone, the enthusiasm of the perfume industry notwithstanding" (p. 329).

Subsequent to the Black and Biron study, Filsinger et al. (1984) had 200 male and female college students rate photographs of a hypothetical male student, called "Paul," on a series of affective rating scales (Osgood et al., 1957). The students also rated their own mood on items from a standardized mood rating scale (Bond and Lader, 1974). The ratings were obtained after the test kits, odorized with either androstenone, methyl anthranilate (a pleasant-smelling agent), skatole (a bad-smelling agent), or no odor, were opened by the students, thereby releasing the odorant. Twenty-two of the subjects (13 men, 9 women) were found unable to smell androstenone at the end of the study, so their data were dropped from consideration.

Individual analyses of variance found no significant group (i.e., odor group), sex, or group by sex interactions for affect ratings, judgments of handsomeness, sexual attractiveness, and strength. A significant group by sex interaction for the activity-passivity rating was found (p < 0.04), reflecting the tendency of men within the androstenone group to rate Paul as more passive and in the methyl anthranilate group to rate him as more active. Women in the androstenone group also reported themselves as less sexy than the women in the other groups. The authors summarize their findings as follows: "The effects in the present study indicated that androstenone had a differential impact on men and women, but the effects did not form a simple and coherent pattern related to sexual activity or excitement. In fact, the effects were somewhat contradictory to the sexual arousal hypothesis, with the female subjects reporting themselves to feel less sexy and the male subjects attributing passivity to the target male" (p. 222).

Influences of Androstenol and Copulin on Social Behavior

Cowley and Brooksbank (1991) had male and female student volunteers wear, from 4:30 in the afternoon until 9:30 the next morning, necklaces odorized with either androstenol, a mixture of aliphatic acids similar to copulin, or a no odor carrier. After removal of the necklaces in the morning, the students were asked to record from memory the number, depth, duration, and type (self-initiated versus other-initiated) of verbal exchanges they had with other persons since the time of waking up. The final study group excluded students who were sharing rooms with one another, resulting in a final sample of 38 men and 38 women.

The influence of wearing the androstenol and fatty acid necklaces on 60 verbal interaction measures was assessed separately for men and women. Each sex recalled having verbally interacted more with individuals of their own sex than of the opposite sex. Nine of 60 analyses comparing the control, androstenol, and fatty acid wearing groups were statistically significant. All significant differences were between women wearing the androstenol necklaces and women wearing the control necklaces in their interactions with men. Thus, the women wearing the androstenol necklaces recalled having more (5.42 versus 3.08), longer (26.7 versus 6.09), and deeper personal involvement (97.4 versus 34.2) interactions with men when the data were considered independently of whether the interaction was initiated by the woman or the man (all ps < 0.025). When the data were analyzed according to who initiated the interaction, self-initiated interactions were recalled as longer (15.1 versus 5.89; p = 0.01), whereas male-initiated interactions were recalled as deeper in personal involvement (54.5 versus 11.9; p = 0.03). The other four significant effects were mostly multiplicative combinations of the number × depth and depth × duration measures.

Like earlier studies, this study has major limitations. First, recall bias may have entered into the study. Second, the units employed in the measures were arbitrary distances on response forms assumed to be isomorphic with the length of interactions, and so on. However, men and women differ in their employment of line lengths in signifying concepts or attributes. Third, specific hypotheses were not articulated. Fourth, given that the necklaces were clearly odorized or not odorized, it is unlikely that the students were blind to the stimuli that they were wearing. Fifth, the data from subjects who had roommates were excluded only after the study was completed, reducing the sample size by 40%. Why weren't the students simply asked to record only those interactions with non-roommates? Sixth, if there is ecological validity to these studies, wouldn't the androstenol from men themselves be a confounding factor, since men wearing the androstenol necklaces would be providing a stronger stimulus than women wearing such necklaces? Seventh, if androstenol is a male-based sex attractant, one might expect men who were odorized with this agent to have more female-initiated exchanges. In fact, the opposite occurred (2.85 male versus 1.31 female initiations, $p < 0.025$). Alternatively, if androstenol was female-based sex attractant, more male-initiated than female-initiated interactions might be expected when women were wearing the androstenol necklaces, but this also did not occur (2.67 versus 2.75, $p = 0.91$). Eighth, the study did not differentiate between dyads in which one or both members of the dyad were wearing odorized necklaces. Ninth, given the large number of analyses that were performed, some would be expected to be significant on the basis of chance alone. Finally, as the authors confide, "Considering the natural setting of the experiment, which inevitably meant that an almost infinite number of variables were not controlled, it is perhaps surprising that any positive findings were obtained" (p. 655). As noted by the authors in the introduction to their paper, "The problem of controlling variables in any social situation are immense, and it is possible to attribute endless reasons for people behaving in the way they do" (p. 647).

Despite the very creative effort of these investigators, this study provides no convincing evidence that either androstenol or copulin serves as a human sex pheromone.

Influences of Androstenol on Mood

In seeming disaccord with a 1984 finding of Filsinger et al. that women felt themselves to be less sexy after being exposed to androstenol, Benton (1982) found no effect of androstenol on women's self-ratings of sexiness. In this study, 18 women rated their feelings on 5 mood scales daily throughout the course of a single menstrual cycle: sexy/unsexy, happy/depressed, lively/lethargic, good tempered/irritable, and aggressive/ submissive. Half of the women rubbed 150 μg androstenol in 70% ethanol on their upper lips each morning, and half ethanol alone. No significant treatment effects were observed for any of the mood scales, although a significant treatment (androstenol/control) by cycle phase interaction appeared for the aggressive/submissive ratings ($p < 0.05$). Twelve one-way ANOVAs on subject group computed for each of the cycle phase designations, which were based neither on hormonal nor basal body temperature, found that during the two days before and the two days after the most mid-cycle day, women in the androstenol group rated themselves as less aggressive and more submissive than women in the non-androstenol group ($ps < 0.05$). When one applies the Bonferroni correction for inflated alpha due to multiple tests, neither of these effects is statistically significant.

Influences of Androstadienone and Estratetraenol on Psychological and Physiological Measures

A number of investigators have examined measures of mood and, in some cases, autonomic nervous system function in subjects before and after inhalation of androstadieneone (AND) or estratetraenol (estra-1,3,5(10),16-tetraen-3-ol) (EST). These steroids are the ones said to alter psychological and physiological responses in a sexually dimorphic manner after infusion into the lumen of the human VNO, as mentioned earlier (Grosser et al., 2000; Monti-Bloch and Grosser, 1991).

In the first of these studies, Jacob and McClintock (2000) applied, in a double-blind paradigm, 250 μM of AND, 250 μM of EST, or a control agent (propylene glycol) to the upper lip of 10 men and 10 women. The subjects then completed questionnaire items from the Profile of Mood States (McNair et al., 2003) and the Addiction Research Center Inventory

(Haertzen, 1974). Difference scores were computed between the control and steroid conditions for each questionnaire item. These scores were then averaged for items loading significantly on three factors derived from a factor analysis: "alertness," "negative-confused mood," and "positive-stimulated mood." Both EST and AND produced larger "positive-stimulated mood" in women than in men, although the latter showed small pre-/post-stimulus decreases under the steroid condition. When the subjects subsequently completed these inventories 2, 4, and 9 hours later, no statistically significant effects emerged. The authors concluded that "this pattern of sex differences in mood does not support a simple sex-specific or sex-exclusive model" (p. 65), in contradiction to the reports of the Utah group (Grosser et al., 2000; Monti-Bloch and Grosser, 1991).

In a second component of this paper, the effects of AND, whose odor was masked with a 1% solution of clove oil added to the propylene glycol diluent base, was evaluated in 31 women. Some additional scales were employed, resulting in two composite factors of "general mood state" and "stimulant/euphoric drug state." AND was found to prevent "the deterioration in general mood which occurred during exposure to the clove oil carrier solution in the laboratory environment" (p. 57). According to these investigators, AND seems to modulate affect, rather than releasing stereotyped behaviors or emotions. On the basis of their findings, Jacob and McClintock conclude that it is "premature to call these steroids [EST and AND] human pheromones" (p. 57).

Subsequently, Jacob et al. (2001) extended this paradigm to include a range of both psychological and psychophysiological test measures in 44 women and 21 men. AND and EST increased hand skin temperature in both sexes. Skin conductance was also increased, but more in women than in men. A similar effect was also noted for "positive mood." Interestingly, the responses of the women, but not the men, were significant only in sessions run by the male tester, implying some social influence of the person doing the testing. These investigators note that "although it is premature to classify these steroids as pheromones, our data suggest that they function as chemosignals that modulate autonomic nervous system tone as well as psychological state" (p. 15).

In another study by this group, Jacob et al. (2002) contrasted psychological responses to AND, androstenol, and muscone (5α-androst-16–3α-ol) in separate test sessions using their standard paradigm. Three factors arose from the factor analysis, which they labeled "Elation-Vigor," "Negative-Confusion," and "Clearheaded-Lucid." Relative to baseline, there was relatively little influence of AND on the three factor-related measures. Interestingly, AND showed less negative change on the "Elation-Vigor" factor items than did androstenol and muscone, and less positive change than androstenol and muscone on the "Clearheaded-Lucid" factor items, an effect the authors characterize as "unique in comparison with those of androstenol and with muscone." The authors reported that androstadienone "prevented both the drop in positive mood and the rise in negative mood that has previously occurred with our experimental protocol" (p. 280).

Studies outside of the Jacob and McClintock laboratory have also reported influences of AND and EST on mood and autonomic nervous system measures. For example, in a study of 30 male and 30 female college students, Bensafi et al. (2004) assessed the influences of AND on various mood scales, as well as on a number of autonomic measures. Rather than applying the agent on the upper lip, they had the subjects take six strong sniffs of the material. Three different groups were used; one group received a high concentration (0.00625 M), another group a low concentration (0.00025 M), and another group the carrier, mineral oil, alone. Relative to men, AND increased positive mood and decreased negative mood in women, with the effect occurring only at the highest stimulus concentration. This agent seemed to have sympathetic-like effects in women and parasympathetic-like effects in men, in that skin conductance was increased in women relative to men, whereas skin temperature was decreased in women relative to men. These effects became larger over the post-exposure period.

More recently, Lundström and Olsson (2005), in a double-blind study employing 37 women, examined the influences of AND in the same formulation as used by the McClintock group on a number of psychological and physiological measures, including ratings of facial attractiveness. Subjects were tested on two test occasions separated by a day—one with AND and the other with the control. Seventeen subjects were tested on both occasions by a man and 20 by a woman. As with the earlier work, the stimuli were applied above the upper lip. Based on a cluster analysis, the psychological responses were collapsed into measures of positive (social, open, relaxed, focused, sensual, energetic, and happy) and negative (heavy, irritated, and down) mood. Relative to the control, androstadie-none was associated with increased positive mood and decreased negative mood when the male was doing the testing. *No AND effects were present when the experimenter was the female. It was also found that under the AND condition women felt significantly more focused only when being tested by the man.* No effects of AND on the measures of attention on facial attractiveness were present. The authors suggest their results imply that social context is important for mood effects induced by AND.

In a similar paradigm, Olsson et al. (2006) determined whether EST exposure influenced the mood of 80 men relative to a control and, if so, if the effect was influenced by the sex of the experimenter. Under the EST condition, the positive mood ratings were less negative than under the control condition, regardless of the sex of the experimenter. *A decrease in negative mood ratings in the presence of a male experimenter and an increase in negative mood in the presence of a female experimenter* were observed, as determined from a significant experimenter sex by stimulus interaction. However, no contrasts were performed and the effects were much larger under the control than under the EST condition. Since only one male and one female experimenter were involved, it is not clear whether the purported sex of experimenter effect was due to sex or to the specific individuals who served as experimenters. If EST were truly a meaningful sex pheromone, one might expect the female experimenter's EST to combine with the EST stimulus to produce a greater effect, which was clearly not the case.

Most of the aforementioned studies represent reasonable attempts to better understand how certain steroids, deemed by some to be human pheromones, influence basic psychological and physiological measures of mood and arousal. Unfortunately, comparisons across studies are hindered by procedural differences and apparent interactions with the sex or personality of the experimenters. The need for relatively large sample sizes and statistical designs where baselines are adjusted to common values to minimize between subject variability suggests that the underlying effects are not large and likely labile. Nonetheless, similar results have been observed across some studies. Overall, these experiments beg the question as to whether odorants other than putative pheromones have similar influences on human behavior and physiology. As will be shown later in this chapter, this seems to be the case.

Influences of Putative Human Pheromones on Functional Brain Imaging Parameters

Following reports that (1) oestra-1,3,5(10),16-tetraen-3yl acetate, a synthetic steroid similar to EST, induces cerebral activation in men at subthreshold concentrations (Sobel et al., 1999) and (2) a putative pheromone receptor gene is expressed in human olfactory mucosa (Rodriguez et al., 2000), Savic et al. (2001) sought to determine, using positron emission tomography (PET), "whether there are compounds that via the nasal mucosa activate the human hypothalamus in a sex-specific mode." These investigators suggested that, if such agents were found, they "would fulfill one important criterium to qualify as candidates for putative pheromones in humans" (p. 661).

In the first and defining study of their series, 12 heterosexual men and 12 heterosexual women served as subjects. Regional cerebral blood flow was measured during passive presentation of EST and AND (Savic et al., 2001). The authors concluded that AND, but not EST, activated the hypothalamus in women, with the center of activation in the preoptic and ventromedial nuclei. In contrast, EST, but not AND, activated the hypothalamus in men, with most activation in the region of the paraventricular and dorsomedial nuclei. In the women, "odor-related brain regions" such as the amygdala and the piriform, orbitofrontal, and insular cortices were activated by EST, but not by AND. In men, neither EST nor AND activated these brain regions although, when the statistical criterion for activation was decreased from $p < 0.05$ to $p < 0.10$, AND produced in men clusters of activation in the right amygdala and piriform cortex, right cerebellum, and right postcentral gyrus. In light of such findings, these authors suggested that "these two steroid compounds may act bimodally, both as pheromones and odors" (Savic et al., 2005, p. 7356).

Subsequent studies by these investigators examined PET activation of brain regions to these steroids in homosexual men and women (Savic et al., 2005; Berglund et al., 2006), as well as non-homosexual male-to-female transsexuals (MFTRs) (Berglund et al., 2008). Additionally, responses to undiluted lavender oil, cedar oil, and eugenol, as well as to a 10% solution of butanol, were assessed. In homosexual men, the anterior hypothalamic region encompassing the preoptic and ventromedial nuclei was activated by AND, but not EST, following the pattern seen in heterosexual women. The lavender oil, cedar oil, eugenol, and butanol activated only the amygdala, piriform, orbitofrontal, and insular cortices. The anterior hypothalamus of the lesbian women was not meaningfully activated by AND, unlike the case of heterosexual women, although this agent did activate the nominal olfactory brain regions. When smelling EST, the lesbian women "partly shared activation of the anterior hypothalamus with heterosexual men" (Berglund et al., 2006, p. 8269). In the transsexuals, the hypothalamus was activated with AND, whereas EST engaged the amygdala and piriform cortex, as occurred in heterosexual women. The investigators concluded, "Because the EST effect was limited, MFTR differed significantly only from male controls, and only for EST-AIR and EST-AND. These data suggest a pattern of activation away from the biological sex, occupying an intermediate position with predominantly female-like features" (Berglund et al., 2008, p. 1900).

These intriguing studies raise the possibility that some volatile agents, most notably steroids, may differentially activate brain regions within or near the hypothalamus in sexually dimorphic ways. It is known that homo- and heterosexual men exhibit hypothalamic activation characteristic of sexual arousal when viewing erotic videos specific to their own respective sexual orientations (Paul et al., 2008), providing a precedent for such differentiation. However, for a number of reasons it is premature to draw a direct analogy. First, one can question whether the chemical stimuli Savic et al. employed should be considered "pheromones" in the first place, regardless of any differential activation of anterior and posterior hypothalamic regions. As mentioned in Chapter 5, the "pheromonal" effects of steroids such as AND are likely conditioned even in pigs, and the rationale for generalizing such agents to humans is weak. It should be pointed out that EST has never been identified on human skin or within human axillary secretions, as is also the case with oestra-1,3,5(10),16-tetraen-3yl acetate. EST is found in the urine of pregnant women, possibly reflecting an intermediary in the biosynthesis of estriol and its epimers (Thysen et al., 1968). Second, the spatial resolution of PET scanning is not high, leading one to question the accuracy of differentiating activity among specific hypothalamic nuclei. Third, since the stimuli of these studies were not equated for perceived intensity, differences in activation could reflect differences in the number of afferent fibers that were activated. Ideally, similarly structured molecules known to influence and not influence endocrine responses should be assessed (Gower et al., 1988). Fourth, in contrast to the findings of Savic et al., the only other study examining the influences of EST on functional parameters in male subjects found activation in 25 to 63% of the subjects in the amygdala and the piriform, orbitofrontal, and insular cortices (Sobel et al., 1999), percentages not dissimilar to those activated by other odorants, including Chanel No 5 perfume (Huh et al., 2008). This work differed from that of Savic et al. in using much lower EST concentrations and employing functional magnetic resonance imaging (fMRI) rather than PET. Moreover, hypothalamic activation to lavender and other non-steroidal odors has been seen in a number of studies, throwing into question the uniqueness of Savic et al.'s findings (Wang et al., 2005). Finally, as they themselves acknowledge, their study had no behavioral or endocrinological correlates, so even if regions of the hypothalamus were differentially activated by the two stimuli in different ways in men and women, it is not clear what this means. Savic (2002) candidly states:

> Although these initial imaging data provide a substrate for the transduction of signals from putative pheromones by humans, they do not prove the existence of human pheromones. Additional studies, including an examination of behavioral aspects in response to pheromone-like substances and evaluations of other candidate compounds are needed to further elucidate this interesting issue. (p. 457)

Proprietary Synthesized Human Sex Pheromones

Cutler and associates have published several studies suggesting that proprietary synthesized male and female pheromones alter the social and sexual behavior of their wearers (Cutler et al., 1998; McCoy and Pitino, 2002; Rako and Friebely, 2004). The design of these studies was similar, save for the use of different putative pheromones and their applications as colognes or perfumes. The paradigm was a two-week-long baseline period during which the subjects used their own individual cologne or perfume, followed by a six-week-long test condition when either ethanol or the putative pheromone plus ethanol was added to the cologne or perfume in a double-blind manner. During both the baseline and treatment periods, daily records of the frequency of six behaviors were recorded: petting/affection/kissing, formal dates (prearranged), informal dates (not arranged before the day of the date), sleeping next to a romantic partner, sexual intercourse, and masturbation. In the case of the women, incidences of approaches by men were recorded. In the case of the men, any change in their experiences with women was noted on a weekly basis.

The first of these studies reported that a pheromone, when added to the cologne of men, increased their frequency of sexual intercourse and sleeping with women (Cutler et al., 1998). A tendency toward increased numbers of self-reported petting/affection/kissing and informal dates, but not masturbation or informal dates, was also found following the use of this agent, which is marketed by Cutler's Athena Institute under the trade name Athena Pheromone 10X. The chemical composition of this mystery agent, which is said to be a synthetic version of a "pheromone naturally secreted by men and described in earlier work (Preti et al., 1987)," was, according to Cutler et al., to be disclosed to the public once the patent process was completed (Cutler et al., 1998). In the second of these studies, McCoy and Pitino (2002) claimed that a proprietary female pheromone, sold under the trade name Athena Pheromone 10:13, increased the frequency of reported sexual intercourse and sleeping next to a partner. An increase in the number of formal dates and petting/affection/kissing was also noted, but no increases in the frequency of male approaches, informal dates, or masturbation were observed.

These two studies suffer from major design flaws. For example, in both cases the placebo and treatment groups were not initially matched on several potentially important variables. In the Cutler et al. study, 6 of the 17 pheromone group participants (35.3%) were married, in contrast to only 3 of the 21 placebo group participants (14.3%). Eight of the placebo group subjects (38.1%) were dating, whereas only 2 of the pheromone group (11.8%) were dating . In the McCoy and Pitino study, 8 of the 19 participants in the pheromone condition (42%) reported they were dating, whereas only one of the 17 participants in the placebo condition (6%) reported such dating. The number of subjects in the pheromone condition who dated "steadily" was 3 (16%), whereas the corresponding number of subjects in the placebo condition was 7 (41%). Relative to the control group, the treatment group reported, *during the baseline pe-*

riod, more male approaches (p < 0.05) and a lower frequency of sleeping next to a romantic partner (p < 0.10). The treatment group was younger (p < 0.10) and shorter in height (p < 0.05) than the placebo group.

Such factors, as well as questionable frequency analyses of increases in behaviors from the baseline to the treatment periods (e.g., the use of χ^2 statistics on cell frequencies as low as one), led Winman (2004) to re-evaluate the data of both of these studies. This author compared the means of the behaviors between the baseline and treatment periods and found no evidence of statistically significant increases in any measure. To the contrary, a statistically significant (p < 0.05) decrease in petting/affection/kissing was noted in the control group of the McCoy and Pitino study during the placebo treatment. Three of the other variables were also lower during placebo treatment (ps < 0.10), namely, sleeping next to a romantic partner, sexual intercourse, and formal dates. A significant decrease over time in the number of sociosexual behaviors during the eight study weeks was observed in the data from the placebo group of the Cutler et al. (1998) study. Winman (2004) concluded,

> It is shown that in neither study is there a statistically significant increase in any of the sociosexual behaviours for the experimental groups. In the control groups of both studies, there are, however, moderate but statistically significant decreases in the corresponding behaviour. Most notably, there is no support in data for the claim that the substances increase the attractiveness of the wearers of the substances to either sex. (p. 697)

A third study on this topic was published by Rako and Friebely (2004) the same year as the Winman critique. This study was similar to that of McCoy and Pitino, with the exception that the 22 experimental and 22 placebo subjects were *postmenopausal* women closely matched on such factors as age, height, weight, and level of education. Like the McCoy and Pitino study, this work was financially supported by the Athena Institute. Hormone replacement therapy was being used by 18 of the women (10 in the pheromone group and 8 in the placebo group, with 17 reportedly using estrogen, 10 using progestin [presumably in conjunction with an estrogen], and 3 testosterone). Aside from the fact that all 3 women who used testosterone fell within the experimental group, the distribution of the hormone user types across the experimental and placebo groups was not indicated.

Unlike what was claimed in the McCoy and Pitino study, Athena Pheromone 10X did not significantly increase the frequency of sexual intercourse, sleeping next to a partner, or formal dates. Congruent with the earlier work, a greater proportion of pheromone (9/22) than placebo (3/22) users reported a higher frequency of petting, kissing, and affection during the treatment period, as determined by χ^2 analysis (one-tailed $p = 0.02$). This effect was weak, however. Thus, if a single subject in the placebo group had increased its performance during the treatment period and the Yates correction for attenuation had been applied, the χ^2 value would not be statistically significant (one-tailed $p = 0.093$).

These authors also sought to determine, using a logistic regression model, which of a number of variables best predicted an "increase versus no increase over baseline during the test period in at least one of the intimate behaviors" (p. 376). None of the dependent measures was examined alone, and the degree of increase or decrease was not assessed. The independent measures evaluated were season (spring, fall), level of education, hormone replacement therapy use, age at menopause, cohabitating or not cohabitating with a male, and pheromone versus placebo use. The only significant predictors were cohabitating versus non-cohabitating with a male ($p = 0.01$) and pheromone versus placebo ($p = 0.04$). In the latter case, the effect was again marginal and it is not clear why no analysis was performed on the independent measures separately. One also wonders why the degree of change from baseline to treatment was not the employed independent measure and why the statistical analyses employed by Winman (2004) were not used.

In light of the discrepant findings among these studies, their funding source, the unidentified nature of the agents involved, the questionable marginal statistical effects, the relatively small number of subjects tested, and the numerous logistic and experimental design problems, one must question whether these putative pheromones have any meaningful influences on human sociosexual behavior.

The Influence of Non-pheromonal Odorants on Emotion and Mood

Are putative pheromones unique in altering human mood, arousal, and sexual behaviors? In this section I describe both animal and human studies showing that a range of odorants not construed as "pheromones" also influence human moods, behaviors, and physiological processes. A number of such agents activate the autonomic nervous system, in some cases increasing and in other cases decreasing arousal. Examples of visual and auditory stimuli that have similar influences on such processes are also briefly mentioned.

Animal Studies

In both rats and mice, vapors of grapefruit oil reportedly increase sympathetic, and decrease parasympathetic, neural activity (Niijima and Nagai, 2003; Tanida et al., 2005, 2006, 2008). Ten-minute-long exposure to such vapors results in significant increases in blood pressure, body temperature, heart rate, and plasma glycerol concentration, as well as the activity of sympathetic nerves that innervate white and brown adipose tissue, the kidney, and the adrenal gland. Such exposure decreases gastric vagal nerve activity. In contrast, exposure to lavender oil has the opposite effects, exciting the parasympathetic gastric nerve and inhibiting the sympathetic nerves that innervate the white and brown adipose tissues and adrenal gland (Shen et al., 2005). The effects of both grapefruit oil and lavender oil on autonomic neural activity are eliminated by electrolytic lesions of the suprachiasmatic nucleus of the hypothalamus and are absent in mice that lack normal circadian rhythms secondary to mutations in genes such as cry 1 and cry 2, implicating this nucleus as critical for their elicitation (Shen et al., 2007; Tanida et al., 2007). Interestingly, auditory stimulation, such as Schumann's *Traeumerei,* but not Chopin's *Etude* or white noise, can also reduce renal sympathetic nerve activity and blood pressure, effects that are also abolished by lesioning the suprachiasmatic nucleus (Nakamura et al., 2007).

Niijima and Nagai (2003) electrically recorded the efferent nerve activity of the central cut end of the sympathetic branch of the nerve innervating the white adipose tissue of the epididymis in anesthetized rats. The sympathetic nervous system is a key regulator of leptin production in white fat. During the session, the rat's head was placed for 10 minutes inside a tilted beaker that contained either blank filter paper, filter paper odorized with grapefruit oil, or filter paper odorized with lemon oil or grapefruit oil, and the frequency of neural activity was recorded before, during, and after the stimulation.. The odorant stimulation, but not the blank stimulation, resulted in a "gradual and remarkable increase in nerve activity that lasted longer than 90 minutes" (p. 1191).

Although there are no direct projections from the olfactory bulb to the suprachiasmatic nucleus, the components of the main olfactory system are intimately associated with circuits that innervate this structure, as measured by retrograde transneuronal tracing using the pseudorabies virus (Krout et al., 2002). In the rat, structures associated with such circuits include the main olfactory bulb, anterior olfactory nucleus, taenia tecta, dorsal endopiriform nucleus, medial amygdaloid nucleus, piriform cortex, and the posteriomedial cortical amygdaloid nuclei.

In common with studies of putative pheromones, these animal studies nonetheless do not differentiate between the influences of the stimuli on the olfactory system and the potential pharmacologic influences from circulatory uptake via the nose or lungs. Moreover, they fail to take into account the possibility that stimulation of the trigeminal nerve may also be involved. Analogous studies in which such factors are controlled are needed to clarify this differentiation.

Human Studies

A substantial literature suggests that odorants not considered to be "pheromones" have significant influences on human emotions, behavior, and autonomic nervous system function. Some of the better-controlled studies are discussed in this section (for reviews, see Moss et al., 2008; Tisserand, 1993; Lawless, 1991; Warrenburg, 2002), along with studies showing that visual and auditory stimuli also significantly influence such factors.

As with the case of mice noted above, certain aromas have been found to increase sympathetic nervous system activity in humans. In one study of 21 young women, for example, such activity was increased by grapefruit oil, as measured using a power spectral analysis of blood pressure fluctuations (Hoppe et al., 2003). Pepper oil, estragon oil, and fennel oil also increased relative sympathetic activity, whereas rose oil and patchouli oil caused a 40% decrease in such activity. A significant change in adrenaline levels was noted only for the inhalation of rose oil (30% decrease). In another study of 26 subjects, exposure to rose, jasmine, and lavender essential oils mitigated the increase in diastolic blood pressure induced by a repetitive handgrip exercise by 24% (Nagai et al., 2000). The odors had no effect on blood pressure elevations produced by a static handgrip exercise said to reflect a lower brainstem reflex, suggesting to the authors that the odor-related reductions reflected central neural processes above the midbrain.

Moss et al. (2008) evaluated the self-reported mood and cognitive performance of 144 students and community volunteers who were randomly assigned to exposure conditions of peppermint, ylang-ylang, and no odor. The stimuli were presented on diffuser pads located under the bench in air- and temperature-controlled testing cubicles and were suprathreshold and approximately of equal strength during each test session. The cognitive testing was done by computer; visual analogue scales were used to assess alertness, calmness, and contentedness before and after the test sessions (Bond and Lader, 1974). Factor analysis was employed to define underlying cognitive factors from a large number of measures. Peppermint was found to significantly increase self-rated alertness, whereas ylang-ylang exposure decreased such alertness ($p = 0.039$). Calmness ratings were higher under the ylang-ylang condition than under both the control ($p < 0.01$) and peppermint ($p < 0.05$) conditions. Factors representing quality of memory, secondary memory, working memory, and memory response speed were all enhanced under the peppermint condition and depressed under the ylang-ylang condition (all $ps < 0.05$).

In an extensive and well-controlled study, Kiecolt-Glaser et al. (2008) performed a randomized double-blind trial to examine the influences of lemon oil, lavender oil, or blank control odor on a number of psychological, psychophysiological, immune, and endocrine measures. Twenty-one women and 35 men participated in 3 separate 6-hour test sessions, one for each odor condition. Half were given no information about the stimuli prior to participation, whereas the other half were given information as the purported relaxing effects of lavender oil and the stimulating effects of lemon oil (so-called priming). The stimuli were presented on cotton balls taped between the nose and upper lip over a surgical tape barrier to prevent percutaneous absorption and were refreshed throughout the session. Lemon oil reliably enhanced positive mood relative to water and lavender oil, as reflected on both the PANAS-positive mood scale (Watson et al., 1988) and three unobtrusive mood assessments (i.e.,

International Affective Picture System valence ratings [Lang et al., 1999], emotional Stroop responses [Mogg et al., 1993], and emotion word use in thought listings). Interestingly, the increase in norepinephrine levels induced by the cold pressor test remained elevated under the lemon, but not the water or lavender, odor condition. Although priming influenced the ratings of the extent to which the subjects expected that their odor would affect their mood and physiological responses, it did not, in fact, influence their actual mood or physiological responses. Surprisingly, none of the odors reliably altered heart rate, blood pressure, IL-6 and IL-10 production, salivary cortisol, skin barrier repair following tape stripping, or pain ratings following the cold pressor stress test.

Ehrlichman and Bastone (1992) had 45 college women rate their mood and complete the Differential Emotions Scale (DES) (Boyle, 1984) while wearing surgical masks impregnated with either a pleasant odor (almond, mint, orange), an unpleasant odor (pyridine), or no odor (water). Those exposed to the pleasant odors reported more positive mood than those exposed to the unpleasant odors. Those in the unpleasant odor group rated themselves as being more disgusted than those in the pleasant odor group. Following exposure to unpleasant odors (butyric acid or thiophene), pleasant odors (almond or maguet), or no odors (control), the subjects rated their feelings on five scales: sleepy-alert, annoyed-pleased, depressed-excited, tense-relaxed, and disgusted-delighted. Ratings were made 1 minute, 14 minutes, and 28 minutes after the odorant presentation. Individuals exposed to the unpleasant odors rated themselves as more tense, annoyed, and disgusted than those exposed to the pleasant odors, a negative affective state that lasted for the entire 28 minutes of testing.

In a study of 56 middle-aged women, Schiffman et al. (1995) examined the influences of 5 different floral fragrances and a placebo on responses to the Profile of Mood States (POMS) questionnaire (McNair et al., 2003). The subjects chose the fragrance they wished to wear at the beginning of the study. The subjects were divided into 4 groups: 14 normally cycling women; 14 non-cycling women receiving estrogen replacement therapy (E); 14 non-cycling women receiving estrogen plus progestin therapy (E + P); and 14 non-cycling women receiving no hormone replacement therapy. In the 12-day-long test sessions, the subjects completed the POMS mid-morning and late afternoon each day after having sprayed themselves with their preferred fragrance. Independent of hormone group, significant fragrance-related increases over baseline and placebo conditions were noted for ratings of vigor and significant fragrance-related decreases for ratings of tension, depression, confusion, and total mood disturbance. In all cases except for vigor, the placebo also differed significantly from the baseline condition, implying a role for suggestion in influencing mood. When hormone group was taken into account, fragrance use was associated with a significant decrement in anger only in the non-menstruating women. Improvement in vigor was found only in the E and E + P groups. Decreased fatigue was found only in the E + P group. Total Mood Disturbance was lower only in the E and E + P groups. Taken together, these data suggest that pleasant-smelling fragrances influence moods in a positive manner and that such influences are most marked in women taking hormone replacement therapy during the postmenopausal period.

Taking an entirely different tack, Graham et al. (2000) found that "masculine" and "feminine" fragrances influenced differential genital responses of women, as measured by changes in vaginal blood volume monitored via a photometric device inserted into the vagina (vaginal pulse amplitude, or VPA). In this study, the fragrances were presented during the follicular and periovulatory phases of their menstrual cycles under control, sexual fantasy, and erotic movie presentation periods. When asked to engage in erotic fantasies during the follicular phase of the cycle, male fragrance produced larger increases in VPA than either a control or a female fragrance. Interestingly, when watching a sexual film during the periovulatory phase of the cycle, a decrease in VAP was noted when female fragrance was present. Seventeen of the 28 subjects (60.7%) were able to identify the male fragrance category, and 21 (75%) the female fragrance category, implying awareness of the type of fragrance employed. No significant effects on ratings of moods, as such, were found. The authors indicated that "whether fragrances are arousing because of their association with past sexual activity, or whether the mechanism involves a direct effect of olfactory stimuli on the brain remains an open question; either explanation would be consistent with our findings, but our data provide no direct support in favor of one of these mechanisms" (p. 83).

The aforementioned studies sampled from a much larger literature on this topic run counter to the notion that axillary odors and putative pheromones uniquely influence measures of arousal and mood. Unfortunately, like the situation with putative pheromones, findings within this literature are not uniform, both sexes are rarely tested in the same test paradigms, and the ranges of odorants examined have been limited. Moreover, stimulus presentation, such as in masks, may lead to habituation and other confounding problems. As is likely the case for most putative pheromones, both psychological and autonomic nervous system responses to odorants are influenced by prior experience. This is clearly shown in a study by Robin et al. (1999). Eugenol (an odorant associated with cements and other agents involved in potentially painful dental procedures), but not vanilla and propionic acid, was found to induce autonomic responses associated with negative emotions (e.g., fear, anger) in persons who fear dental procedures, based on bad experiences, but not in ones who do not.

It is critical to point out that olfaction, regardless of whether activated by putative pheromones or traditional odorants, is not the only sensory system that influences autonomic, psychological, and endocrinological responses of humans. For example, higher color temperature work environments improve mental activity and subjective mood, increase sympathetic autonomic tone, and ameliorate to some degree drowsiness, fatigue, and daytime sleepiness (Hoffmann et al., 2008; Noguchi and Sakaguchi, 1999; Mukae and Sato, 1992). An hour-long exposure to 800 lux of light in the early morning has been shown to significantly increase morning salivary cortisol levels (Scheer and Buijs, 1999). Music enhances arousal and mood (Husain et al., 2002; Thompson et al., 2001) and improves cognition and mood in patients recovering from brain damage (Sarkamo et al., 2008). In light of these and the aforementioned observations, is it reasonable to conclude that pheromones exist that uniquely influence physiological and psychological mood states in humans?

Human Menstrual Synchrony Pheromones

In a highly publicized 1971 *Nature* paper that spawned the rat estrous synchrony work described in Chapter 6, the menstrual cycles of close friends or roommates who live together in a dormitory were said to synchronize over time (McClintock, 1971), that is, the onset of their period of menstrual bleeding became more close over a six-month period. Subsequently, a number of studies reported similar synchrony (e.g., Graham and McGrew, 1980; Goldman and Schneider, 1987; Little et al., 1989; Matteo, 1987; Quadagno et al., 1981; Skandhan et al., 1979; Weller and Weller, 1992, 1993a, 1993b, 1995b, 1997a; Weller et al., 1995, 1999a), although some reported negative findings (e.g., Cepicky et al., 1996; Jarett, 1984; Strassmann, 1997; Trevathan et al., 1993; Weller and Weller, 1995a, b; Wilson et al., 1991). Since the publication of the defining study 35 years ago, however, not a single report has appeared in which the chemical identification of the alleged menstrual cycle synchrony pheromone has been made. Importantly, as described in detail below, a sizable scientific literature has since appeared suggesting that menstrual synchrony, like estrous synchrony, is a questionable phenomenon, having no viable evolutionary basis and likely reflecting statistical artifact (Arden and Dye, 1998; Schank, 1997, 2001b, 2000b; Strassmann, 1999, 1997; Wilson, 1987, 1992; Schank, 2006; Ziomkiewicz, 2006; Yang and Schank, 2006).

Does Menstrual Synchrony Exist?

Wilson (1987) carefully assessed the criteria employed up to the mid-1980s that were used to define menstrual synchrony, and concluded that, on the basis of statistical issues, synchrony had not been demonstrated in any of these studies (i.e., McClintock, 1971; Graham and McGrew, 1980; Russell et al., 1980; Quadagno et al., 1981; Preti et al., 1986). Wilson pointed out that the only apparent difference between studies reporting and not reporting such synchrony was that the latter included persons with irregular menstrual cycles and that omitting persons with irregular cycles biased the results toward synchrony. Three sources of error, based on the original McClintock method, were identified by Wilson, with the latter two having a particular propensity to bias toward an erroneous conclusion of synchrony: I. An implicit assumption that differences between menses onsets of randomly paired subjects vary randomly over consecutive onsets; II. An incorrect procedure for determining the initial onset of absolute difference between subjects; and III. Exclusion of subject data on the basis of not having the number of onsets specified by the research design. Error I includes the failure to take into account that approximately 50% of paired cycles of unequal length will show a tendency to synchronize by chance when relatively few cycles are evaluated. Error II reflects two factors: first, the fact that an incorrect onset difference (which only occurs for the initial onset calculations in McClintock's method) is always greater than a correct onset difference (which occurs for subsequent onset calculations), increasing the mean onset absolute difference and erroneously leading to what seems to be synchrony in subsequent onsets; second, an incorrect onset difference reverses the direction of change between the consecutive onset differences of a pair. This occurs because the subject with the earliest recorded onset has the latest recorded onset after the correction. Error III biases samples toward showing menstrual synchrony by reducing dispersion in final onset absolute differences, a common phenomenon in studies finding evidence of menstrual synchrony.

Error II was explained in a simple manner in Cecil Adams's column, *The Straight Dope,* in *The Chicago Reader* newspaper (Adams, 2002). Assume a menstrual cycle study starts on October 1. The first study subject reports a 28-day-cycle with an onset of menses on September 27, another onset on October 25, and a third on November 22. The second study subject, with a 30-day cycle, reports a menses onset on October 5 and another on November 4. Using McClintock's calculation in which only cycle onsets are recorded within the study period, 20 days separated the 2 menses onset dates (October 25 versus October 5) and 18 days separated the second pair of menses onsets (November 4 versus November 22). This calculation would suggest that the two cycles are synchronizing when, in fact, they were 8 days apart to begin with (September 27 versus October 5) and actually diverging.

Such methodological criticisms led a number of investigators to revamp the procedure they used to determine menstrual synchrony. Weller and Weller, for example, employed a "last months only" (LMO) paradigm to assess menstrual synchrony in subsequent work (e.g., Weller and Weller, 1997a, 1993a, b, 1998, 1997a, 1997b; Weller et al., 1999a, b). The aim of the LMO procedure is to determine whether a significant degree of synchrony exists in a sample of women who have been together for some period of time relative to either expected frequencies of onset differences based upon random onset occurrences or to random reassignment of new "random pairs" of women from the sample. Weller and

Weller (1997b) indicate that this procedure "avoids the problems posed by cycle variability and irregular cycles, for unlike the previous methods, this procedure does not measure initial onset differences and employs Wilson's modification for examining the relationship between menstrual occurrences. It simply assumes that if menstrual synchrony occurs in the sample, this procedure would demonstrate it, if the women had lived together for a sufficiently long period of time. The use of an appropriate null model ('expected distribution') or a random comparison group controls for much of the difficulties introduced by cycle variability and the occurrence of irregular cycles" (p. 120). These authors stress, however, that it would be preferable to employ several procedures to establish synchrony longitudinally, and that synchrony would be clearly demonstrated if all of the procedures proved positive.

Unfortunately, the LMO approach also has its limitations. As Weller and Weller (1997b) themselves note, studies that have sought to examine menstrual synchrony over time have a relatively low ($\sim$ 50%) participation rate, reflecting issues related to volunteering, accurate record keeping, and provision of requested data (e.g., return of menstrual calendars). Additional substantive concerns have also been raised by others regarding the LMO method (Arden and Dye, 1998; Schank, 2000b, 2001b). Schank (2000b), for example, performed a computer simulation employing assumptions underlying the LMO analysis, and found that cycle variability introduced a systematic bias toward synchrony; the greater variability in the simulated cycle distribution, the greater the bias. He presents data suggesting that even if cycle onsets are completely randomly related, the LMO synchrony measurement would lead to data distributions skewed toward synchrony, and that such skewing occurs "in a way that is qualitatively and quantitatively like the actual data distributions they [Weller and Weller] report" (p. 842). Subsequently, he pointed out that an unexplained pattern of a systematic increase in *asynchrony* occurred over time (i.e., across days or months) in all studies employing the LMO approach, regardless of the grouping variable (e.g., close friends, roommates, sisters, families) (Schank, 2001b). He notes that "without some theoretical reason for increasing asynchrony after initial observation, one should instead expect increased synchrony (McClintock, 1971; Schank & McClintock, 1992) or small random fluctuations around the initial mean onset difference in succeeding months if a persistent state of synchrony exists (i.e., the women in this study [of Bedouin families] had been living together and were assumed to be in a state of synchrony; A. Weller & Weller, 1997)" (p. 3). He goes on to point out, in all eight groups contained within the five studies he reviewed, that "if it is assumed that small fluctuations about the initial mean onset difference are random, then there should be a 50–50 chance that each group becomes numerically more or less synchronous. The probability that all eight groups should increase in *asynchrony* by chance is only $.5^8 = .004$" (p. 4).

Wilson and Schank are not the only critics of menstrual synchrony. Strassmann (1997) questioned not only the validity of menstrual synchrony per se but the assumption made by some that such synchrony is biologically adaptive. She points out that in most preindustrialized societies, specifically natural-fertility societies that presumably reflect the norm for much of human evolution, pregnancy and lactation, not menstrual cycling, takes up the majority of the female's reproductive years. Unlike in industrialized societies, pregnancy occurs in the early teenage years and there is little attempt to control fertility in a parity-dependent manner. In a long-term prospective study of the Dogon of Mali, Strassmann examined 477 untruncated menstrual cycles from 58 women over a 2-year period (Strassmann, 1997). In this society of millet farmers, strict taboos require menstruating women to be segregated at night in special huts. Straussman, through a nightly census of the women present in the huts (736 days), was able to obtain information about the onset of menses without interviews and errors in recall or reporting. Hormonal analyses confirmed the accuracy of these measures as an indicator of the onset of menstruation. Compared to American women who have, on average, more than 400 menstruations in their lifetimes, Dogon women have an average of 128 menstruations.

Over the course of the study, the proportion of women cycling on a given day was about 25%. Approximately 16% were pregnant, 29% were in lactational amenorrhea, and 31% were postmenopausal. On any given day, subfecund women were most common among the cycling women, and conception usually occurred for the most fecund women on one of their first postpartum ovulations, resulting in their dropping out of the pool of regularly menstruating women. Employing statistical procedures that overcame the problems pointed out by Wilson (1987) in calculating menstrual synchrony, she found no evidence for synchrony of the cycling women who habitually ate and worked together or who lived with a particular lineage of related males. Moreover, no evidence for synchrony was observed in any of the remaining cycling women. She concluded her discussion section as follows: "Given the paucity of evidence, it is surprising that belief in menstrual synchrony is so widespread. I suggest that this belief arises, in part, from a popular misconception about how far apart one would expect the menstrual onsets of two women to be by chance alone" (p. 128).

Strassmann subsequently elaborated on this point elsewhere:

Popular belief in menstrual synchrony stems from a misperception about how far apart menstrual onsets should be for two women whose onsets are independent. Given a cycle length of 28 days (not the rule—but an example), the maximum that two women can be out of phase is 14 days. On average, the onsets will be 7 days apart. Fully half the time they should be even closer (Wilson, 1992; Strassmann, 1997). Given that menstruation often lasts 5 days, it is not surprising that friends commonly experience overlapping menses, which is taken as personal confirmation of menstrual synchrony. (Straussman, 1999, p. 128)

The aforementioned studies cast significant doubt on whether menstrual synchrony itself is a real phenomenon. If it is a true biological phenomenon, one might expect reproductive synchrony to be more focused on ovulation than menses, making the latter measure an imprecise index of synchrony, particularly when anovulatory cycles are included (Weller and Weller, 1997b). If, in the seemingly unlikely event that menstrual synchrony exists in some groups of subjects under very specific circumstances, the question arises as to whether pheromones are involved in the synchronization process. As noted in the following section, it would seem that evidence for pheromonal mediation of putative synchrony is weak and fraught with procedural issues (Doty, 1981; Wilson, 1992, 1987; Whitten, 1999; Schank, 2002, 2006).

If Menstrual Synchrony Exists, What Evidence Is There That Pheromones Are Involved?

It should be noted from the outset that many studies have suggested that factors other than pheromones influence putative menstrual synchrony.

Most such studies also suffer from the methodological issues raised by Wilson, Schank, and others, throwing into question the validity of their claims of synchrony and perhaps providing a basis for why so many variables have been associated with the putative synchrony (Wilson, 1987, 1992, 1993; Schank, 2000b, 2001b, 2002). Little et al. (1989) reported that synchrony occurred within a month in 76 women who resided in 12 housing units at a university regardless of whether they lived in the same residence, suggesting that common environmental conditions played a role in producing synchrony. Matteo (1987) examined synchrony over a 3-month period in women working together as graduate students (n = 10) or in an all-female university department (n = 10), a typing pool (n = 8), an emergency room (n = 6), or a recovery room (n = 7). Women who experienced high levels of anxiety and job stress were reportedly less synchronized than those with low levels of anxiety and stress, and "menstrual synchrony occurs in occupational settings in which job interdependency is equal to or greater than job stress" (p. 473). Jarett (1984), while not finding statistically significant evidence of synchrony among 80 roommate pairs in 2 Catholic colleges, reported that "movement toward menstrual synchrony was predicted by the use of sanitary napkins, long menstrual flow, a low score on the affiliation scale of the Personality Research Form, a high score on the social recognition scale of the PRF, and not mentioning stress in guessing the purpose of the study" (p. 25). Goldman and Schneider (1987) assessed synchrony in 140 female volunteers in 3 college residence halls and found synchronization scores to be high among friend pairs, next highest among roommates that were not listed as friend pairs, and lowest among randomly paired pairs. Friends that were similar on the neuroticism scale of the Eysenck Personality Inventory, an index of emotional reactivity, were said to have "synchronized very closely," leading these authors to conclude: "The close synchronization of friends who were similar in personality implies an interactive model based on both social contact and personality similarity. None of these findings challenge the explanation for synchronization based on pheromones, but factors beyond time spent together influence synchronization" (p. 249).

The first specific claim of a demonstration of pheromone-induced synchronization of menses was that of Russell et al. (1980). In this experiment, axillary secretions were collected on gauze pads taped under the arm of a woman who had a history of regular 28-day menstrual cycles and a "previous experience of 'driving' another woman's menstrual cycle on three separate occasions, over three consecutive years; i.e., a friend had become synchronous with her when they roomed together in the summer and desynchronized when they moved apart in the fall" (p. 737). The gauze pads were then removed, cut into 4 squares, combined with 4 drops of 70% alcohol, and frozen in dry ice. Following thawing, the material from appropriate phases of the cycle was rubbed on the upper lips of 5 women, 3 times a week, for 4 months. A control group of 6 women were similarly rubbed with gauze pads that had received only an alcohol treatment. A mean pretreatment difference of 9.3 days was observed between the day of the onset of the donor's menses and that of the experimental subjects. The average difference decreased to 3.4 days after the fourth month of treatments, suggesting a trend toward menstrual synchrony. The authors concluded that "the data indicate that odors from one woman may influence the menstrual cycle of another and that these odors can be collected from the underarm area, stored as frozen samples, for at least short periods, and placed on another woman. Further, the experiment supports the theory that odor is a communicative element in human menstrual synchrony, and that at least a rudimentary form of olfactory control of the hormonal system is occurring in humans in a similar fashion to that found in other mammals" (p. 738).

Careful scrutiny of this study, however, raises some critical questions. First, the study was not performed in either a single- or double-blind fashion. Indeed, the woman who donated the samples (the second author of the paper) also acted as one of the two female experimenters who rubbed the stimuli on the subjects (Doty, 1981). In addition to potentially providing subtle social cues that might affect the experiment's outcome, she would presumably be confounding the experiment with a second source of pheromones, namely, those on her person during her interaction with the subjects. Second, the purpose of the study was explained to each subject. Whether this alone could influence its outcome is not clear, although social stimuli can be zeitgebers for some human biological rhythms (including rhythms of certain hormones) and have the potential to override even the dominant mammalian zeitgeber, light (Wever, 1979; Rusak and Zucker, 1979). Third, Wilson (1992) examined the data of this study in light of the three errors previously mentioned, indicating that this study

> shows evidence of all three errors: The number of synchronous cases is too few to be statistically significant (Error I), one of the four synchronous cases has an incorrect initial onset difference which, when corrected, causes the initial mean onset difference to be greater than the final mean onset difference (Error II), and one or more subjects may have withdrawn from the experiment because their cycle behavior was not meeting the expectations of the investigators (Error III). I conclude that Russell et al. (1980) did not demonstrate menstrual synchrony in subjects treated with axillary extract from a female donor. (p. 577)

In a study similar to that of Russell et al.'s, Preti et al. (1986) sought to correct a number of the problems of the Russell et al. study by using a double-blind procedure and not informing subjects about the purpose of the experiment until after the study was completed. The 19 subjects were selected from a larger number on the basis of self-reports of regular cycles (29.5 ± 3 days), presumably minimizing the potential adverse influences of highly irregular cycles. A solution of alcohol and axillary secretions obtained from cotton pads previously worn in the axillae during "a convenient 6- to 9-hr period" of 4 female donors was applied to the upper lips of 10 of these subjects 3 times a week for three complete menstrual cycles. The stimuli employed reflected 3-day segments of the cycles of all 4 subjects from which they were collected, resulting in a set of "donor cycle" stimuli whose midpoints consisted of donor cycle days 2, 5, 8, 11, 14, 17, 20, 23, 26, and 29. The extracts were said to have been applied in 22- to 25-day intervals. Eight of the 10 subjects in the experimental group reportedly synchronized with the extract treatment schedules after 2 complete cycles, whereas only 3 of 9 of the control women did so. The authors state that "this study represents the first systematically designed, prospectively conducted, double-blind research in humans to attempt to manipulate the menstrual cycle with female-derived secretions. In this experiment naturally occurring 29.5 ± 3 day cycles could be modulated with repeated applications of extract at a 22 to 25 day interval. This study establishes phenomena in humans which are analogous to previously demonstrated olfactory/reproductive relationships in nonhuman mammals" (pp. 480–481).

In a critical assessment of this study, Wilson (1987) reanalyzed Preti et al.'s data and found that "the apparent synchrony in menses onsets in the axillary extract sample is explained on the bases of (a) chance variations, (b) mathematical properties of co-cycling menses onsets, (c) features of the experimental design, and (d) failure to follow the experimental protocol, or calculation errors, or both." In his reanalysis, Wilson found 20 instances, equally divided between the experimental and control group data, where the cycle length of the treatment application fell outside of the 22- to 25-day range stipulated in the experimental protocol. Thus, in the extract sample, the donor's cycle was found to be greater than 25 days in 9 instances, and less than 22 days in one instance, a point later acknowledged by Preti (1987). Since most of Wilson's analyses are beyond the scope of this presentation, I simply quote his summary:

> In summary, the equal distribution of five preovulatory and five postovulatory cases in the extract sample is due to chance. Eight of these cases are shown [in Table 1] as having decreased absolute onset differences between the first and third onsets of the subjects and donor. The decreases in the four preovulatory cases, including two cases in which the subject had constant cycle lengths, are interpreted as a product of the experimental design, the mathematical properties of co-cycling menses onsets, and chance variations. The decreases in the four postovulatory cases, including one case with constant cycle lengths, are interpreted as the result of "errors" in the cycle lengths of the treatment applications. If all of the treatment cycles were in the 22- to 25-day range specified by the experimental protocol, the extract sample would have the characteristics of a sample of randomly paired subjects. No evidence in this experiment suggests that the 29.5 ± 3 day cycles of the subjects in the extract sample were modulated by the applications of the female axillary extract or that humans have phenomena analogous to olfactory/reproductive relationships demonstrated in nonhuman mammals. (pp. 537–538)

A more widely publicized study claiming axillary influences on menstrual synchrony was that of Stern and McClintock, which appeared in *Nature* in 1998. These authors state the following: "We found that odourless compounds from the armpits of women in the late follicular phase of their menstrual cycles accelerated the preovulatory surge of luteinizing hormone of recipient women and shortened their menstrual cycles. Axillary compounds from the same donors which were collected later in the menstrual cycle (at ovulation) had the opposite effect: they delayed the luteinizing-hormone surge of the recipients and lengthened their menstrual cycles. By showing in a fully controlled experiment that the timing of ovulation can be manipulated, this study provides definitive evidence of human pheromones" (p. 177).

In this study, 9 donor women wore cotton pads in their axillae for at least 8 hours after bathing. The pads were collected by the experimenters on a daily basis, and urinary LH (luteinizing hormone), along with other information (e.g., menses, basal body temperature), was used to "classify each pad as containing compounds produced during the follicular phase (2 to 4 days before the onset of the LH surge) or the ovulatory phase (the day of the LH surge onset and the 2 subsequent days)" (p. 179). The pads were cut into 4 sections for distribution to different subjects, treated with 70% isopropyl alcohol, and stored at $-80°$ C until use. The subjects were studied for one initial cycle without exposure to the axillary stimuli. After thawing, the secretions on the pad were then applied to the upper lips of the subjects on a daily basis during the next 4 consecutive cycles. Ten of the subjects received material from the follicular phase each day for 2 menstrual cycles and then material from the ovulatory phase for the next 2 cycles. The reverse was the case for the other 10 subjects. The donors served as a control group, receiving only the 70% alcohol carrier above their upper lip on each day. The authors noted that "in addition, because the two-day change in menstrual cycle length (expected for the initial study) is substantially less than individual variation in cycle length typical for this age group, we created within-subjects controls by measuring the effect on the menstrual cycle in terms of a change in length from each individual subject's cycle preceding each condition" (p. 179).

Stern and McClintock reported that the stimuli from the follicular phase produced shorter cycles than the stimuli from the ovulatory phase (-1.7 ± 0.9 days versus $+1.4 \pm 0.4$ days). Unlike earlier studies of menstrual synchrony by McClintock et al., this effect occurred within the first cycle. The carrier had no effect on cycle lengths of the controls. The authors noted that "in five of the cycles, women had mid-cycle nasal congestion, which could have prevented their exposure to pheromones; including these cycles in the analysis made the results slightly less robust (follicular compounds: -1.4 ± 0.9 days; ovulatory compounds: $+1.4 \pm$

0.5 days; ANOVA: follicular versus ovulatory compounds F $(1,18)$ = 4.32, P $\leq$ 0.05; cycle 1 versus cycle 2 of exposure (not significant, NS); order of presentation (NS); all alternations between factors were not significant)" (p. 177).

In a second element of the study, the investigators sought to "determine the specific mechanism of pheromone action." To do so, they utilized the LH and progesterone data to establish the follicular and luteal cycle phases. Armed with this information they "traced all the changes caused by the pheromones presented in our study to the follicular phase. For the menses and luteal phases, the distributions during the pheromone and control conditions were the same (indicated by overlapping log-survivor curves). Only the follicular phase was regulated, shortened by follicular compounds and lengthened by ovulatory compounds, suggesting that these ovarian-dependent pheromones have opposite effects on the recipient's ovulation by differentially altering the rate of follicular maturation or hormonal threshold for triggering the LH surge" (p. 178). These workers concluded that "this experiment confirms the coupled oscillator model of menstrual synchrony and refocuses attention on the ovarian-dependent pheromones that regulate ovulation, producing either synchrony, asynchrony or cycle stabilization within a social group, namely two distinct pheromones, produced at different times of the cycle, which phase-advance or phase-delay the preovulatory LH surge" (p. 178).

Aside from any attempt to isolate the chemical agent(s) that might be considered the pheromones, the Stern and McClintock study, like the work on synchrony itself, has been criticized on numerous methodological and conceptual grounds. Schank (2006), for example, points out that in their analysis of the five cycles, rather than subtracting the onset dates of the first cycle from that of the following four cycles, these authors subtracted the onset dates of cycle 1 from those of cycles 2 and 3, and the onset dates of cycle 3 from those of cycles 4 and 5. In effect, cycle 3, in which axillary odor was being applied, was treated as a baseline period when, in fact, it was a treatment period. In his critique, Schank provides examples of why such an analysis is flawed. Moreover he demonstrates how random data sets drawn from a truncated normal distribution with the means and standard deviations reported by Stern and McClintock become statistically significant only after being transformed using the flawed McClintock analysis procedure.

Strassmann (1999) similarly was critical of this study, pointing out that the authors disregarded all of the previous methodological critiques that undermine the validity of synchrony research based upon the McClintock methodology and questioned the robustness of the statistics used to support the finding:

The conclusion that a change in cycle lengths of the subjects was caused by a pheromone, rather than by the well-documented variation in cycle length in women (Treloar et al., 1967; Harlow and Zeger, 1991), requires inordinate confidence in the biological importance of a P value of borderline statistical significance ($P \leq 0.055$). From the data presented it is unclear whether the assumption of a normal distribution was justified. Moreover, in view of the small sample size, the entire effect might have been due to just one or two subjects who had undue leverage. Additional questions are raised by the following statement (Stern and McClintock, 1998): "Any condition preventing exposure to the compounds, such as nasal congestion anytime during the mid-cycle period from 3 days before to 2 days after the preovulatory LH, could weaken the effect. We analyzed the data taking this into account." It would be useful to know what a priori criteria were employed in making such adjustments, and whether the data analysis part of the project was done blind. In the absence of a theoretical reason for expecting menstrual synchrony to be a feature of human reproductive biology, and until a cycle-altering pheromone has been chemically isolated, it would appear that skepticism is warranted. (p. 580)

Whitten (1999), one of the pioneers of mouse pheromone research, also questioned the validity of the Stern and McClintock study. Among his concerns was one pointed out by Strassmann; namely, the fact that "each group has an apparent outlier favourable to the model: one of -14 comprises 25% of the total shortening, whereas that of $+12$ makes up 22% of the increase. Excluding these two outliers would abolish the claim of significance" (p. 232). His major point of concern, however, was as follows:

My main criticism of the study is the use of the value of single first cycles, receiving carrier-only treatment, to derive the data analyzed. Such single observations have no within-subject variance and the irregular statistical manoeuvre of converting all 20 observations to zero masks any between-subject variance and provides an illusory zero baseline with indeterminate confidence limits. Carrier-only treatments should have been distributed throughout this long experiment to give a balanced crossover design with three treatments (carrier, follicular and ovulatory) and two or more complete replications to confer confidence limits to the baseline observations, thus making comparisons valid. (p. 232)

Whitten concluded his critique as follows: "I am not convinced of the validity of the coupled-oscillator model derived from rat studies. I also question the 'definitive evidence' that pheromones regulate human ovarian function because, if these exist, their characterization will require large, carefully designed experiments, a controlled social and physical environment, and a clearly defined endpoint measured in hours" (p. 232).

The other element of the Stern and McClintock study, the changing of the timing of the LH surge, has received little attention, although Shinohara et al. (2001) theorized that if male pheromones can influence the timing of ovulation in female mammals, it is not too farfetched to conclude that female pheromones might be able to do the same. These authors collected axillary extracts analogous to those collected by Stern and McClintock from 5 women 2 to 4 days before the LH surge (follicular phase) and on the day of the LH surge or one to 2 days after the surge (ovulatory phase). Using 20 subjects randomly divided into 3 groups (follicular phase recipients [n = 8], ovulatory phase recipients [n = 7], and isopropyl alcohol control recipients [n = 5]), stimuli were applied to the subjects' upper lips after an initial 4-hour period of no treatment. Blood samples were collected repeatedly at 30-minute intervals throughout the initial period and a subsequent 4-hour stimulation period. The frequency of the LH pulses was increased slightly by follicular phase compounds and decreased by ovulatory phase compounds, but not by the alcohol. In an earlier study, these authors had reported that the purported pheromone androstenol (see the following section) decreased the frequency of LH pulses in a manner similar to ovulatory phase axillary secretions (Shinohara et al., 2000). In light of such findings, Shinohara et al. argued that pheromones play a role in the modulation of the timing of ovulation by changing the frequency of pulsatile LH secretion.

A close assessment of the two papers by Shinohara et al. (2001, 2000) raises concerns about their validity on several grounds, as described by

Schank (2006). First, as in the case of the Stern and McClintock study, the sample sizes were quite small and, importantly, the authors did not monitor or control for factors known to influence LH surges, such as menstrual cycle stage (Filicori et al., 1998). Second, the significance values of the paired t-tests that were reported in these small samples would seem to require that all of the women responded in the same direction to the putative pheromone, in contrast with the 68% response rate noted by Stern and McClintock. Third, the interpulse intervals of the androstenol example depicted in their 2000 paper are more than 5 standard deviations from the reported mean interpulse interval of the group receiving the androstenol. As Schank notes, the "reported standard error is mathematically impossible to calculate with this extreme outlier when n = 6" (p. 464). Fourth, again as noted by Schank, the data suggest that the 2000 and 2001 papers were not truly independent. Thus, the intra-assay coefficients of variation for the LH analyses were the same for the two studies, as was the age range and the number of women in the control groups. In Shinohara et al.'s 2000 study, the control mean before stimulus application was 59 minutes (SEM = 4.9) between LH pulses and after application it was 58 minutes (SEM = 4.7), whereas in their 2001 study the control means before and after application were nearly 5 minutes less than in the 2000 study (i.e., 54 and 53.9), even though exactly the same standard errors were reported. When Schank calculated the means for the frequency of pulses, he found them to agree with the interpulse intervals for the control conditions across the two studies, although the standard deviations were different. If these discrepancies are due to error, Schank suggests they are unlikely typographical and seem to bias the 2001 paper in the direction of the authors' conclusions.

Even if it were the case that some biological secretions marginally alter LH activity, should such alterations be considered being mediated by pheromones? Several artificial odors or irritants also can alter LH activity. For example, exposure of men and women in the luteal (not follicular) phases of their cycles to 50 ppm toluene for 3 hours influences LH secretion, although such exposure does not result in abnormal LH or FSH secretion profiles (Luderer et al., 1999). As described in Chapter 6, in some animals the LH surge can be classically conditioned to a number of stimuli, including odors. Thus, an arbitrary odor, after minimal conditioning, will ultimately elicit a male LH response, even in the absence of odor previously associated with a female (Graham and Desjardins, 1980). Should such arbitrary odors be considered pheromones?

Breastfeeding Human Pheromones

Jacob et al. (2004) have reported that substances contained in the axillae and nipple secretions from breastfeeding mothers increase the variability in menstrual cycle length of nulliparous women receiving these compounds. Although they refer to such agents as chemosignals, they conclude the abstract of their paper as follows: "Because compounds from lactating women and their infants modulated the ovarian cycles of women, as is seen in other mammals, they have the potential to function as pheromones, regulating fertility within groups of women" (p. 422).

In this study of 47 women, the first menstrual cycle was a baseline cycle in which all subjects were exposed to pads moistened with a control carrier substance. During the next 2 cycles, 27 of the women continued to receive the control pads and 25 received pads that had been worn next to the axillae and breasts of lactating women for 8 hours. During the 3 months of the study, each subject was instructed to wipe the pads on the skin above their upper lips at least 4 times a day. Each woman returned to the laboratory twice a week to receive a new set of pad-containing vials. The investigators who interacted with the subjects at that time were blind as to whether the provided pads were control or odorized pads, although the subjects were asked to report on each of these laboratory occasions whether they smelled anything on the pads. Basal body temperature (BBT) was recorded each morning, as was a recording of the evening cervical mucus characteristics, evidence of menses, sexual activity, and ratings of sexual motivation and desire. A week prior to the midcycle they tested their urine every evening for the surge in LH using an ovulation kit. Urinary progesterone was measured periodically after the LH surge was evident. This information provided a means for distinguishing the follicular and luteal phases of the cycle.

This study, like the earlier studies of this group, has come under close scrutiny and has been criticized on methodological grounds. Schank (2006), for example, points out that that these authors plotted the cycle lengths for the baseline and second cycle for the experimental and control groups, but did not plot out the cycle length data for the third cycle in the same fashion. While the investigators performed regressions among the menstrual cycle lengths of cycle 1, cycle 2, and cycle 3 for both the odor exposed and control exposed groups, they provided neither descriptive nor statistical analyses of the cycle variability per se. To address the variability issue statistically, Schank (2006) reconstructed the cycle-length data sets from the data points provided in the figures of the publication. Although he encountered inconsistencies between the data represented in the figures and data reported in the text, he was able to obtain a reasonable reconstruction of the data set. Using a formal test of unequal variance, he found no evidence for a statistically significant difference in variability between the cycle lengths of the subjects exposed to the control pads and those exposed to the axillary and breast stimuli.

As with the case of human menstrual synchrony, this phenomenon seems questionable on the basis of statistical grounds. This, along with the fact that no "pheromone" has been isolated which would induce the effect, makes one question whether this can reasonably be considered a pheromone-mediated phenomenon.

EIGHT

Implications

··

They hunted till darkness came on, but they found
Not a button, or feather, or mark,
By which they could tell that they stood on the ground
Where the Baker had met with the Snark.

In the midst of the word he was trying to say,
In the midst of his laughter and glee,
He had softly and suddenly vanished away—
For the Snark *was* a Boojum, you see.

From the Eighth Fit, "The Vanishing," in Lewis Carroll's
The Hunting of the Snark

I began this book with a reference to Lewis Carroll's poem, *The Hunting of the Snark*, which chronicles the "impossible voyage of an improbable crew to find an inconceivable creature." I end this book by elaborating further on the problems with the pheromone concept and how scientists can extract themselves from the chaos of misinformation that arises from its use. As described later in this chapter, this dichotomous concept, with its emphasis on innateness, simple chemical stimuli, and species specificity, is found wanting even when applied to social insects. Unfortunately, like astrology, the lack of consensual definitions and formal postulates makes the pheromone concept less than amenable to proof or disproof. This is particularly true when a multitude of adjectives can be added to the term *pheromone* to make the concept fit any desired perspective and when no attempts are made to isolate chemical agents that can be subjected to empirical test.

The lack of consensus as to what defines a pheromone, as outlined in

Chapter 2, along with the tantalizing concept of an externally secreted hormone, opened the floodgates of the popular press to sensationalize the pheromone concept. As can be seen when one searches the Internet for the term *pheromone,* millions of sites suggest that pheromones are chemicals that have dramatic influences on human social and sexual behaviors. In fact, pheromone-laden products are now a substantial element of the multibillion-dollar personal care products industry. Some go so far as to believe that pheromones are all around us, influencing our every move. Kodis et al. (1998) state, in *Love Scents,* the following:

> Pheromones are odourless molecules that are produced in the body and enter the world by wafting off the skin. They also float up from the recesses of the sweat glands and linger in strands of hair. Each unleashed pheromone molecule is packed with information about your sexual desires, your level of aggression, the attributes of your immune system, and more. Every pheromone carries your one-of-a-kind chemical "signature," which is as unique as the swirls of your fingerprint. Pheromones tell you about your neighbor, your best friend, your coworkers, the man who reads your electric meter, the person who sits next to you on the bus. (p. 12)

Such perversion reflects the lack of any solid scientific or operational meaning of the pheromone concept or its signification by the term *pheromone.* When a concept reaches this state of affairs, why not abandon it? I am often asked, "If you do away with the term *pheromone,* what term should be used in its place?" My answer is none, although admittedly some alternative terms are less provocative and problematic when communication, per se, is the referent (e.g., *semiochemical* or *chemosignal*). Before 1960, studies of the influences of biologically derived chemicals on mammalian behavior and endocrinology did quite well without applying a value-laden generic label to the stimuli. It would seem more prudent to describe relationships between organisms and their environment in operational terms. As exemplified by my 2003 *Handbook of Olfaction and Gustation* (Doty, 2003b), the term *pheromone* can be omitted from the scientific lexicon without any loss of information. Occam's razor can be applied and operationalism evoked. As expressed by Albert Einstein (1934): "The supreme goal of all theory is to make the irreducible basic elements as simple and as few as possible without having to surrender the adequate representation of a single datum of experience" (p. 165).

Sorensen and Stacey (1999), while maintaining the belief that natural selection is focused on *specific* chemicals, point out the practical problems with the pheromone concept in terrestrial vertebrates—problems that would seem insurmountable if one accepts the premises of their perspective:

> Presumably, chemical stimuli are predisposed to function as social signals because they are ubiquitous and discriminated with great sensitivity and specificity. Because of this specificity, organisms detect only a portion of the myriad compounds surrounding them. Pheromonal systems have proven challenging to study because it has been difficult to predict which of the many chemicals released by organisms might have pheromonal activity. Terrestrial vertebrates appear to have evolved, repeatedly and independently, a variety of sex pheromones with no clear common precursors. Few of their sex pheromones have been identified and no general theoretical framework has emerged to systematically address either the diversity of sex pheromone systems or the evolutionary processes that might have created them. (p. 16)

A more parsimonious explanation of these difficulties is that such agents rarely, if ever, exist in terrestrial vertebrates. In mammals, single chemicals of innate origin, which act alone to induce complex non-learned behavioral and endocrine responses are, for all intent and purposes, nonexistent. This is reflected by the fact that nearly a half century has gone by and no specific chemicals that could be construed as unique determinants of behavioral or endocrine responses have been isolated. Complex sets of chemicals, not single chemicals, seem to be the norm for communication of important social information critical for mammalian evolution, such as individual identity (Johnston and Jernigan, 1994; Beynon and Hurst, 2004). From the perspective of information transfer, the metabolic cost of making numerous small molecules is less than that of making a few complex molecules, particularly when existing metabolites can be employed. Varying elements of a complex stimulus array allow for nearly infinite levels of information available for communication, particularly when learning can intervene. Also antithetical to the pheromone concept is the fact that the brain cannot be divorced from the communicative process, and that information transfer typically occurs in a noisy environment with dynamic fluctuations and temporal inconsistencies in the sensory arrays that require top-down management. The dangers of simplistic stimulus-based concepts for chemical communication are aptly pointed out by Wilson and Stevenson (2006) in relation to the main olfactory system:

> Although it is of some importance to understand what characteristics of the stimulus yield particular sensations, this is likely to tell us very little about olfactory *perception*. At its most basic, the search for consistent relationships between stimulus and sensation is predicated on the idea that the stimulus produces a set response by activating a particular receptor(s), the response being a sensation that is, all other things being equal (e.g., anosmia, adaptation), *the same* in all participants at all times. However, this simple and apparently useful starting point is undermined when consideration is given to what the olfactory system actually needs to accomplish to "smell." The most difficult problem is how it identifies a biologically significant odor from the array of other odors present at any one particular time. A stimulus-response system is of little value in this respect, because all the system registers is what is presented to it at the receptor level; no attempt can be made with such a system to select a particular pattern of stimulation over that of any other. A related problem also arises. Biologically significant odors are typically not single pure chemicals; rather, they are complex mixtures composed of tens or hundreds of volatile substances (Maarse, 1991). Thus, the problem facing the system is even more complex than one first imagines, because the system has to select not just one biologically relevant stimulus but a pattern of stimuli that may themselves change over time and place and that co-occur against a constantly shifting background of stimulation... the implications arising from these two points are fatal to any theory of olfaction that relies solely on a stimulus-response mechanism. (p. 17)

A core problem with the pheromone concept is its division of stimuli into two classes—pheromonal and non-pheromonal. Conceptually this is dangerous, since mutually exclusive categories cannot share attributes or features and preclude the existence of multiple classes or continua. Such dichotomies inaccurately represent diversity and limit the range of possible options, forcing adherents to fit any number of phenomena into one or the other class. The attempts to define pheromones on the basis of such bipartite categories as innate versus learned, single versus multiple, conspecific versus heterospecific, olfactory versus vomeronasal, volatile versus nonvolatile, and hypothalamic versus non-hypothalamic seem doomed from the start. Not only are a number of these distinctions difficult or impossible to make, but there are 64 possible combinations of even these 6 categories. After deciding whether agents are pheromones, mammalian pheromonologists have been forced to add modifiers to the term *pheromone* to make it more compatible with reality. Thus, the original term *releasing pheromone* soon morphed into such terms as *signaling pheromone, behavioral pheromone, modulatory pheromone,* or *informer pheromone* to reflect the reality that behaviors are rarely "released" by chemicals in mammals. The term *primer pheromone* also took on new identities, most typically morphing into the generic term *pheromone* without any modifiers or qualifications. This was done primarily by investigators who were either unaware of behaviors that had already been ascribed to the term or who felt it unnecessary or unwise to acknowledge their existence. Dozens of types of pheromones have since emerged, as exemplified by the subheadings outlined in Chapters 5 and 6.

The tendency to name pheromones in relation to specific behaviors or endocrine effects, as illustrated by these subheadings, is reminiscent of the assignment in the late nineteenth century of instincts to nearly every behavior imaginable. By the early twentieth century, hundreds of human instincts had been chronicled by instinct theorists, although today the term only rarely appears in psychology textbooks. When the criteria were critically evaluated that differentiated instinctive from acquired responses, only one was found to be universally applicable—that instincts are unlearned (Beach, 1955).

Perhaps it is not coincidental that, as with instincts, learning seems to be the prime defining factor used in attempts to differentiate between pheromones and non-pheromones. In this regard, the pheromone concept seems to be a very close cousin to the instinct concept. Modern instinct theorists, such as Lorenz (1970), expanded Craig's (1918) concept that behavior patterns are separable into appetitive and consummatory components, the latter reflecting the instinct. Lorenz suggested that a continuum exists across species, represented at one end by species with extremely specialized instinctive behavior patterns and at the other end by species in which "the entire motor activity developed to perform an adaptive function is left open to purposive control. The greater the intelligent capacities of a given animal species, the greater the extent to which the goal can be left open to purposive behaviour, until eventually the typically instinctive conclusive act of the behaviour pattern is restricted to an emotional or motivational response to a situation" (p. 289). Within this mode of thinking, a good example of the admixture of purposive behavior and an instinctual behavior is scent marking. In the case of dogs, there is an inherent predisposition for males to lift their legs to urinate and for females to squat to urinate (Figure 4.1). However, these responses are not invariant. In addition to being influenced by the intrauterine hormonal milieu, they are largely under cognitive control. Thus, a male dog recognizes the odor of another male and only then directs his micturition response to the stimulus representing that male. As illustrated by the mouse micturition patterns shown in Figure 3.1, social factors determine when and where scent marking is put into effect.

Even though experience largely determines the responses of mammals to most chemical stimuli, the basic machinery necessary for chemosensation is inherited. Leaving aside the complexity of mixtures, it should be emphasized that even single chemicals have multiple elements that invariably activate limited but diverse sets of receptor-bearing cells within the olfactory or vomeronasal epithelia. These sets of cells, in turn, dictate spatially segregated glomerular maps within the primary or accessory olfactory bulbs (see, e.g., Johnson and Leon, 2007). Isomorphism of such topography continues, to some degree, to higher brain regions. However, neural topography is not the primary determinant of the information upon which responses are based, despite being a necessary substrate. In other words, the hardware is necessary for, but does not inextricably determine, the software. In the case of the main olfactory system, successive transformations of the neural program occur at numerous levels, beginning with the receptors and extending to the olfactory bulb, olfactory cortices, and multiple brain regions that interrelate with these structures. The meaning of the information to the organism is established at higher levels where past experiences are encoded.

While much of this book has emphasized the important role of learning in influencing chemically mediated behaviors, one must accept the fact that the physical and physiological processes that allow for such learning are themselves under genetic control. Moreover, some aspects of odor communication are undoubtedly immutable to modification by learning. As pointed out by Mayr (1974), however, it is illogical to contrast genetics against experience, a problem that frequently rears its head as a component of the pheromone concept:

> The history of behavioral biology is a history of controversies... Much of the argument concerning animal behavior dealt with the question, "How much is *innate* and how much is *acquired* through experience?" The trouble with this terminological dichotomy is that innate refers to the genotype, and consequently neither term is the exact opposite of the other one. (p. 651)

One way out of this dilemma is to infer, as suggested by Mayr, the existence of neural programs that are open or closed to change by experience and that are translations of underlying genetic programs. Ethologists and evolutionary biologists would then focus on the evolutionary history of the genetic programs, whereas physiologists and psychologists would focus more on the open or closed neural programs. Mayr's concept provides a template for the co-involvement of both experience and genetics. Such approach also makes it easy to envision a continuum, dictated by genetics, of neural programs that are represented at one extreme by those totally closed to experiential modification and at the other extreme by those totally open to such modification. In the case of chemical influences on behavior and physiology, varying degrees of modification are then possible, some occurring *in utero* (e.g., the influences of exposure to chemicals in the mother's diet that alter the offspring's later food choices) and some at other stages of development (e.g., the learning of species odors during nursing). Importantly, genetics and epigenetics, as well as immunological influences, can be fully included for consideration.

Critics may point to successes in the isolation of insect pheromones as a reason for continuing the search for mammalian pheromones. While such successes are well established, it is now apparent that many responses to putative insect pheromones, particularly those of social insects, are also significantly influenced by learning, context, and physiological state. Although mammalian behavior is more complex than insect behavior and, hence, seemingly less amenable to the tenets of the pheromone concept, many chemically mediated insect behaviors fail to comport well with this concept. The entomologist Hölldobler (1999) states the following:

> In the early stages of the study of chemical communication in animals, scientists assumed that in insects behavioral responses are released by single

chemical substances, whereas in vertebrates and particularly in mammals, chemical signals are complex blends of substances, mediating inter-individual recognition and interactions. However, most insect semiochemicals have proven to consist of several compounds, whereby different components of a complex pheromone mixture may have different effects on the receiver. Thus, with respect to the sophistication of their chemical communication systems, vertebrates and insects do not differ greatly. (p. 129)

The fact that the stimulus-based pheromone concept diverts attention away from the multimodal communication that typifies most interactions among conspecific mammals has also not gone unrecognized in the study of invertebrates. Hölldobler (1999) notes, in Gibsonian fashion, the following about ants and other social insects:

I think we have underestimated the complexity of communication signals in ants, having focused our analyses on one or the other sensory channel through which the signals are perceived and processed. The lesson we have learned from studies of chemical signals, namely that insect semiochemicals have proven to be complex mixtures, and single-compound pheromones are actually quite rare, have now to be extended to other modalities, and greater attention has to be paid to multimodal combinations of single components. Certain components function as straightforward releasers, others as subtle modulators of motivational states of the receivers. We have relatively little problem when considering motivational or emotional states in mammals affecting the individual's readiness to send or perceive communication signals, but we are still quite hesitant to investigate the role of motivation in social insect communication. Yet those of us studying communication in social insects are fully aware of the fact that the behaviors of the signal sender, as well as that of the signal receiver, depend strongly on the motivational states of the individuals, although we cannot precisely measure motivational states. (pp. 139–140)

It is clear, as demonstrated throughout this volume, that the pheromone concept, which is not tightly tethered to consensually validated definitions, oversimplifies the nature of mammalian chemical communication and is, frankly, unnecessary, non-parsimonious, and non-operational. This stimulus-based concept promotes the questionable idea of singularity of function of molecules in chemical communication, detracts from the appreciation of learning and cognitive processes, brings to mind externally secreted hormones, implies uniqueness of chemical stimuli in altering a number of endocrine processes, and leads to unfruitful chemical searches for magic bullets. Importantly, this concept is based largely upon inbred strains of rodents in highly artificial situations where interspecific relationships, such as occurs in natural populations, are absent. In light of such problems, where do we go from here?

As mentioned earlier in this chapter, a reasonable first step would be to employ operational terms in describing the relationship between chemical stimuli and behavioral or endocrine responses. For example, rather than saying that "pheromone-containing male urine was placed on the snout of a female mouse, inducing subsequent uterine growth," one would say, "male urine was placed on the snout of the female mouse, inducing subsequent uterine growth." The second step would be to recognize that single chemicals or small sets of chemicals unlikely dictate most behavioral or physiological responses to natural chemical stimuli. This is not to denigrate the importance of understanding receptor ligand interactions, but to promote diversity as to the factors that commonly determine behavioral and physiological responses. This includes developmental and experiential factors, as well as potential interactive influences from nonchemical sensory modalities. Third, while one must guard against unbridled anthropomorphism, eliminating such thinking from the development of logical theories throws the baby out with the bathwater. While the emphasis must remain upon quantifiable processes if science is to progress, assuming continuity of anatomy and physiology across mammalian species while denying any commonality of emotional or mental processes is an obvious oxymoron. Although other mammals may not think in the same manner as we think, their responses to the environment have undergone many of the same evolutionary pressures. Fourth, under the assumption that rodents and other mammals have cognition and feelings, investigators should be sensitive to the nature and ramifications of their experimental manipulations. For example, some chemically induced endocrine changes observed in the laboratory may well reflect influences from highly abnormal sensory stimuli that have quite a different meaning to a mouse than to our own species. In humans, stress can accelerate the onset of puberty and can induce blockage of implantation. One wonders what would happen to pubertal timing or implantation of prepubescent girls if, without warning, abrupt subjugation to a strange middle-aged adult male or his pictures, letters, or videotapes was made on an hourly or daily basis throughout the course of the pre-pubertal or post-insemination periods in a *deus ex machina* fashion. Finally, focusing on species-specific communication, as implied in the pheromone concept, has been sorely limiting, as many of the selection pressures for the development of scent marking and scent glands reflect influences outside of species members. Such influences can come from predators as well as closely related forms found in regions of sympatry. For example, the Indian desert gerbil, *Meriones hurrianae,* responds to the midventral sebaceous scent gland secretions of the sympatric nocturnal gerbil, *Tatera indica,* by significantly shortening its home range and conspicuously increasing its own scent-marking activity (Idris and Prakash, 1986). Ignoring interspecific interactions where evolutionary processes are strongly at work has likely limited our understanding of the evolution of scent glands, marking behaviors, and chemical communication in general (Moore, 1965; Doty, 1972, 1973; Doty and Kart, 1972). Hopefully, awareness of this problem will result in a resurgence of interest in studying non-laboratory animals in their natural habitats.

A primary goal of this book is to explicate the problems with the pheromone concept when applied to mammals. It shows that attempts to identify and isolate mammalian pheromones have been snark hunts and the pheromone concept adds little value to understanding chemical communication. The point is made that some of the phenomena attributed to pheromones, such as menstrual synchrony, are themselves suspect in terms of their existence. Despite such problems, the pheromone concept seems to continue to serve as an iconic elixir for many scientists in their exploration of the mysterious world of relations between animals and their social and physical environments. The multiple attempts to redefine, rather than abandon, the pheromone concept is a referendum on its compelling nature. In light of such compulsion, I am under no illusion that this book will reverse the beliefs of many scientists that pheromones mediate or explain most mammalian behaviors. However, I do hope that this contribution will facilitate discussion about the nature of chemical communication and will aid in countering malignant forms of reductionism that fail to see the forest for the trees.

From "The Vanishing." Illustration by Henry Holiday from Lewis Carroll's
The Hunting of the Snark: An Agony in Eight Fits
(Macmillan, New York, 1891).

Notes

....

Chapter One. Introduction

1. It could be argued that our own species has the tendency to underestimate the cognitive component of the perceptual processes of other animals, laying bare our centricities and desire for reductionism.

Chapter Two. What Is a Mammalian Pheromone?

1. In 1939, Hartmann and Schartau coined the term *gamone* for all substances involved in sexual reproduction in lower forms, including endohormones, ectohormones, and membrane-bound recognition factors. This term has proved too general and impractical to gain widespread acceptance (Maier and Müller, 1987).

2. Further divisions were sought to overcome the problems with the pheromone concept in insects. Karlson and Butenandt (1959) suggested, for example, that the term *pheromone* be reserved for chemicals with more specific biochemical effects (presumably hormone-like) and that the term *telomone* be used for chemicals that induced responses via sense organs. Brown (1968) proposed the terms *allomone* and *kairomone* for situations where *interspecies* communication occurs. Allomones were those chemicals adaptively favorable to the transmitter and kairomones ones adaptively favorable to the receiver. As with pheromones, these chemicals could be produced or acquired by the organism, encompassing the notion that dietary factors can influence their chemistry.

3. Although one might argue that the mutual benefit derives from survival of the group, defining a priori such benefit seems difficult to operationalize.

4. Bronson goes on to point out that the murine nervous system not only contains many more neurons than that of an insect, but differs significantly in terms of degree of encephalization, the numbers of associative neurons, and the flexibility afforded to the mediated behaviors. He indicates that while the releas-

ing pheromone concept is a valuable tool for describing the "often relatively simple, stimulus-response systems" of insects, it is questionable for mice and other mammals (p. 123). In light of such issues, Bronson suggested in 1968 that the term *signaling* should replace the term *releaser*. In 1973, Whitten and Champlin suggested that *behavioral* should serve as the substitute. Subsequent investigators suggested replacing the term *releasing pheromones* with such terms as *social odors* (Brown, 1979), *homeochemic substances* (Martin, 1980), or *semiochemicals* (Albone, 1984).

5. There are cases where the vomeronasal organ does appear to be preferentially stimulated by some natural secretions in mammals. For example, following exposure to estrous urine odors, sexually experienced male rats evidence an increase in Fos immunoreactivity within the accessory olfactory system (Kippin et al., 2003; Novotny et al., 1990; Singer, 1991). An artificial almond odor, conditioned to the act of copulation, did not demonstrate such activity within the accessory system, but did demonstrate such activity within the main olfactory pathways.

6. Paradoxically, as Nodari et al. (2008) point out, only a few volatiles, major histocompatibility complex (MHC) peptide ligands, and major urinary proteins (MUPs) found among the many known or suspected constituents of mouse urine have been shown to stimulate murine vomeronasal sensory neurons. At saturated concentrations, these stimuli increase the firing of fewer than 15% of such neurons, in contrast to the 30 to 40% of neurons activated by highly diluted mouse urine (Holy et al., 2000). This led Nodari et al. to screen multiple neurons within the vomeronasal epithelium for responsiveness to a range of fractions of female mouse urine identified by chromatography and other techniques. In their chemical analyses, many sulfated steroids were found to be present in female, but not male, mouse urine. Using 31 synthetic sulfated steroids and a similar number of stimuli comprised mostly of purported VNO ligands, the sulfated steroids were found to trigger responses 30-fold more frequently than the previously reported ligands, suggesting such steroids are important VNO stimulants. Nevertheless, the median (range) percent of active electrode channels in their multi-electrode array system that was activated by the 31 steroids was 2.4% (0–19) for the male and 2.9% (0–17.2) for the female epithelia. Moreover, the responses to the sulfated steroids were similar in male and female epithelia ($r = 0.55$, $p < 0.001$), a curious finding in light of their observation that such steroids are present only in female urine. A key finding of this study was that urine from stressed animals contains high levels of a possible metabolite of corticosterone, in accord with the idea that such scent sources provide information directly related to physiological status rather than simply serving as deliberate signals released by animals for communication.

1. Opioids, in conjunction with noradrenegic pathways, appear to play a significant role in memory consolidation and in suppressing neonate fear behavior in neonatal mice by modulating the activity of the amygdala (Roth et al., 2006).

2. Learning of odors in the early postpartum period is not confined to the neonate. Recognition of the odors of the offspring by mothers is also learned during this time. Although this has been demonstrated in humans (Schaal et al., 1980), the classic examples of this phenomenon, which some have postulated as being mediated by pheromones, come from studies of hoofed mammals (e.g., deer, cattle, and sheep) (Dunn et al., 1987; Price et al., 1984). In these species, the mother typically learns the odor of her offspring immediately following parturition when she licks and cleans off the afterbirth and placenta. Both the main and accessory olfactory systems are involved (Booth and Katz, 2000). However, an artificial odor can serve equally well as the natural odor as the learned stimulus (Poindron et al., 1993). Thus, while a traditional means of getting ewes to accept alien lambs is to impart the odor of a familiar lamb onto the alien (Price et al., 1998), artificial odors that have been made familiar to the ewe have been found to be as effective as natural odors in determining such acceptance (Price et al., 1998; Alexander and Stevens, 1985). Interestingly, the hormonal environment of late gestation sets up the stage for the mother's period of olfactory learning and acceptance—a period triggered by mechanical stimulation of the vagina and cervix during parturition (Kendrick et al., 1992).

3. Such effects are not confined to affiliative social or mating preferences. House mice (*Mus musculus*) and deer mice (*Peromyscus maniculatus*) reared in the presence of both species' odors prove to be more successful in heterospecific agonistic encounters than their conspecific counterparts that have been reared only with their own species' odors (Stark and Hazlett, 1972).

4. While much has been made of odor-related dissortative mating preferences in mice related to genes of the major histocompatability complex (MHC) (Yamazaki et al., 1998; Beauchamp et al., 1985; Yamazaki et al., 1976; Brown and MacDonald, 1985; Doty, 1986b), genes at other loci are also involved in establishing cues employed in individual identity and cross-fostering, and diet and bacterial factors can attenuate and even override such genetic predispositions (Penn and Potts, 1998; Schellinck and Brown, 1999; Sherborne et al., 2007). For example, Yamazaki et al. (1988) demonstrated in mice whose genetic differences were only within the MHC complex that the preference for B6-H-2^k males to mate with B6-J-2^b females, and the preference for B6-J-2^b males to mate with B6-H-2^k females, was reversed when the mouse pups were cross-fostered by the

opposite H-2 haplotype. More recently, Hurst et al. (2005) found that MHC-associated odors did not stimulate countermarking in a territorial situation, suggesting that associative learning was involved in territorial recognition and that "MHC-associated odours were neither necessary nor sufficient for scent owner recognition" (p. 715). In a semi-natural field study, these investigators found no evidence of avoidance of mates with the same MHC genotype in wild mice when genome-wide similarity was controlled (Sherborne et al., 2007), although avoidance of mates with the same major urinary protein (MUP) haplotype was noted.

5. It is of interest that a number of bird species selectively employ, from hundreds of alternatives, aromatic plants in nest development, and that they work to maintain an odor-laden environment over the period of time when the offspring are raised (Petit et al., 2002). The preferred plant species contain chemicals known to be antibacterial, antiviral, fungicidal, and insecticidal and/or insect repellent, including linalool, camphor, limonene, and eucalyptol. When experimenters artificially decrease the aromatic environment, the birds increase the use of aromatic plants, whereas the reverse occurs when experimenters increase the intensity of the aromatic environment. The degree to which this phenomenon depends upon experience with the aromatic agents on the part of the nestlings is not known.

6. It is important to note that prior experience not only influences responses to odors, but can alter sensory processing in a way that enhances discrimination and subsequent perceptual learning. A case in point is a recent human study in which optical isomers (mirror-image molecules), unable to be initially discerned by odor from one another, became discernable after several trials during which one of the isomers was associated with an electric shock (Li et al., 2008). This rapid change in performance was mirrored by divergence of ensemble activity patterns in the piriform cortex, as measured by functional magnetic resonance imaging (fMRI). This effect was not due to heightened arousal or attention, and suggests that emotional experiences with an odor can significantly alter the ability to discern that odor from other closely related odors.

7. Recognition of not only one's own species, but of individual members within that species, is critical for the survival of most so-called higher organisms, as without such recognition natural selection would be greatly compromised. While the role of odors in species and individual recognition varies from group to group, it is of interest that members of every terrestrial mammalian order tested to date can recognize conspecific individuals by their odors (Goyens and Noirot, 1975; Carr et al., 1976). The importance of individual recognition, at least in social species, is reiterated by Clark (1982), who appropriately states that "individual recognition and its effects on behavior may be the single major basis for structuring mammalian and avian social relations. Its importance is easily appre-

ciated in dominance relations, mother-offspring recognition, kin-directed behavior, and mate recognition, as familiar examples" (p. 1153).

8. Other studies suggest that adult odor exposures can also influence nesting preferences. Thus, adult female mice of the SEC1Re/J strain (SEC), when exposed to the odor and sounds of C57BL6/J (C57) males for 7 days during the immediate post-weaning period, exhibit a preference for bedding odors from C57 males over that males of their own strain when tested in estrus at 120 days of age (Albonetti and D'udine, 1986). This preference is independent of whether they are fostered by a SEC or a C57 dam until weaning.

Chapter Four. Scent Marking

1. Subordinate mice reportedly are more likely to spend less time in scent-marked areas than dominant mice, implying that the odor from dominant males may be to some degree aversive (Jones and Nowell, 1989). In many cases this is likely learned in the process of becoming subordinate, although small mice raised in isolation are more likely to avoid scent-marked areas than large mice raised in isolation, conceivably reflecting differences in androgen titer or self-awareness of their own competitive ability (Gosling and Roberts, 2001).

2. Exploratory behavior does not seem to fall into this category. For example, sexually experienced male guinea pigs have been noted to essentially ignore, in a home-cage preference test, urine from a human male, as well as urine from a non-receptive female *Galea musteloides* (Beauchamp, 1973).

Chapter Five. The Elusive Snarks: Case Studies of Nonhuman Mammalian "Releasing" Pheromones

1. In adult males, but not prepubertal ones, exposure to female hamster vaginal secretions (FHVS) has been associated with increases in testosterone levels, although mature sexually experienced males exhibit a higher baseline testosterone level, as well as a larger secretion-related increase in testosterone from the baseline, than mature sexually inexperienced ones (Romeo et al., 1998; Macrides et al., 1974). The extent to which this phenomenon reflects conditioning is not known. As noted elsewhere in this book, arbitrary odors can be conditioned to elicit endocrine responses. Interestingly, FHVS-induced Fos-immunoreactivity in forebrain nuclei associated with male reproductive behavior (e.g., medial nucleus of the amygdala) does not differ between sexually naïve prepubertal and sexually naïve adult male hamsters (Romeo et al., 1998), suggesting that the FHVS-influenced mating behavior of the adult does not reflect maturation in the responsiveness of the afferent or central nervous system (CNS) reproductive structures. The mating-induced pattern of neuronal activation in these regions, as measured by c-*fos*

expression, is influenced by numerous factors in mature males, including (1) previous sexual experience in the testing environment, (2) the presence of FHVS, and (c) the number of ejaculations achieved in sexual encounters (Kollack-Walker and Newman, 1997).

2. Five of the tests were conducted using 10-minute test periods, and the rest using 5-minute periods. The 5-minute periods were used once it became clear that the behavior on the part of the male was initiated and maintained early in the session and that 10-minute periods were not needed.

3. There is some evidence that the nature of the sexual experience and other factors may alter the saliency of the preference (Le Magnen, 1952a; Taylor and Dewsbury, 1990; Brown, 1977; Clegg and Williams, 1983; Kendrick, 1975). A similar phenomenon has been demonstrated in the house mouse, although strain differences appear to be present (Hayashi and Kimura, 1976; Rose and Drickamer, 1975). In some species the development of a preference for estrous over diestrous odors seems complex (Taylor and Dewsbury, 1988). In prairie voles, animals that have a more monogamous social structure than many rodents, the preference for estrous odor does not develop in males receiving sexual experience through monogamous cohabitation or in males housed with two females. However, males exposed to both estrous and diestrous females, and males housed with other males and females in a semi-naturalistic setting, develop this preference.

4. Approximately 40 kHz vocalizations are also found once mounting occurs, interspersed among the 70 kHz vocalizations, which also decline across this period. Their function is presumably related to copulation per se (White et al., 1998).

5. In accord with this concept is the finding by Dixon (1982) that increased aggression directed toward mice injected with diazepam was seemingly due to changes in the odor of their urine.

6. The odorant could conceivably be the milk itself (since fresh rabbit's milk is equally preferred in a choice test to the exposed abdomen of a lactating female), although this is believed unlikely.

Chapter Six. The Elusive Snarks: Case Studies of Nonhuman Mammalian "Priming" Pheromones

1. One might argue that if stress is the basis of the Bruce effect, then the increase in prolactin levels known to occur in some mice as a result of acute stress would provide a source of endogenous prolactin that would mitigate the pregnancy block. One reason why this does not occur may be the fact that progesterone and other steroids serve to profoundly modify prolactin responses in response to stress (Jahn and Deis, 1986).

2. Estrogen also rises in response to chronic stress (MacNiven et al., 1992).

3. One study reports that duration of exposure required to make a stud male familiar to the female is 3 to 4.5 hours (Rosser and Keverne, 1985).

4. Some data do not support the familiarization hypothesis. For example, Lott and Hopwood (1972) reported that if the stud is removed from the female within 3 hours of mating, pregnancy block from a strange male is less likely to occur than if he remains with the female for 24 hours or longer, suggesting that exposure to the stud "sensitized" the female to subsequent pregnancy block.

5. A more common explanation of this phenomenon is that the inseminated female requires time to learn the characteristics of the stud male (Bronson, 1976a). Circumstantial support for the hypothesis that it reflects habituation comes from rats, where continuous exposure of a female to the same male produces fewer estrous cycles than successive exposures to different males (Cooper, Purvis, and Haynes, 1972; deCatanzaro, Baptista, and Spironello-Vella, 2001).

6. It is of interest that exposure of rat dams and their litters to male rat urine from days 14 to 29 postpartum delays the second lactational estrus by 2 days (relative to distilled water exposed controls), and results in a 1-day *delay* in vaginal opening and first estrus in the female offspring. Female offspring also exhibit a shorter first estrous cycle and tend to produce larger litters (Schank and Alberts, 2000). Such observations suggest that exposure to male urine increases, rather than decreases, sexual maturation in rats, in contrast to mice, although transgenerational effects appear to be present.

7. Early work suggested that numerous fractions of peptide-like materials each contained biological activity, as determined from uterine weight increases. According to Novotny et al. (1999b), such fractions, despite repeated purification, lyphilizations, and long-term purging, continued to yield numerous volatile organic molecules and each possessed an odor reminiscent of mouse urine. However, in all attempts to chemically isolate the puberty-accelerating pheromone, the isolates have rarely, if ever, exhibited the biological activity of the whole urine (e.g., Vandenbergh, Whisett, and Lombardi, 1975).

8. Not all stressful stimuli need to activate adrenal responses. For example, social isolation, long considered to be a "stressor," can influence a number of behaviors (e.g., agonistic behavior, scent marking, emotionality) independent of clear changes in plasma corticosterone levels or adrenal gland weight (Spencer et al., 1973). Yoshimura (1980) found prolonged isolation increased scent marking of male gerbils, a behavior known to be largely androgen dependent. The cholinergic antagonist, scopolamine, however, suppressed such marking behavior independent of any measurable changes in central acetylcholinesterase or choline acetyltransferase activity.

9. The nature of the post-handling housing may, however, influence this effect. Thus, while housing female rats in small cages after such handling on a daily

basis from birth to 30 days of age advances the time of vaginal opening, housing rats in groups of 10 in a large cage within an enriched environment during this period delays the age of pubertal opening (Swanson et al., 1983). Perhaps paradoxically, in both the rat and the mouse, brief periods of daily handling in the early postnatal period (e.g., 3–15 minutes) decrease the magnitude of adult behavioral and endocrine responses to stress.

10. Initially, maturation of the ovarian follicles is observed, presumably from increased estrogen production. Subsequently, massive stroma luteinization occurs, reflecting the effects of progesterone, luteinizing hormone, and luteotrophic hormone. Associated with the second phase are (1) degenerative changes in the follicles and ovaries; (2) increased merocrine and holocrine secretion in the adenohypophysis that accompanies increased gonadotrophic function; (3) changes in neuro-secretion within the hypothalamus (e.g., in supraoptic nucleus); and (4) enlargement of both the supraoptic and paraventricular hypothalamic nuclei, accompanied by a decrease in the amount of secretory material, occurring maximally about 10 days after the onset of the stress regimen. Increases in corticotrophic and thyrophic hormones occur as well.

11. Although all-female groupings are rare in nature, this is not the case in animal husbandry. As the result of changes in animal welfare legislation within the European Union, pigs are now group housed rather than individually housed. This has led to the finding of impaired reproduction in group-housed sows compared to individually housed sows, reducing pregnancy rates and the number of piglets born per litter (Kongsted, 2004).

Chapter Seven. Human Pheromones

1. The question as to whether humans have vaginal pheromones that denote the time of optimal receptivity has also been raised. Unlike many mammals, humans do not have a "heat" period, and evidence for greater mid-cycle sexual activity is equivocal (Hill, 1988). Nonetheless, odorous volatiles from the vagina fluctuate to some extent across the phases of the menstrual cycle, arising largely from bacterial action upon (1) vulvar secretions from sebaceous, sweat, and Bartholin's and Skene's glands; (2) cervical mucus; (3) endometrial and oviductal fluids; (4) transexudate through the vaginal walls; and (5) exfoliated cells of the vaginal mucosa (Huggins and Preti, 1976). In 1975, my colleagues and I performed a study in which 37 men and 41 women judged the odor intensity and pleasantness of vaginal secretions obtained from tampons worn by four women at various phases of their menstrual cycle (Doty et al., 1975). The mid-cycle phase was determined from basal body temperature charts. The subjects were not informed as to the nature of the stimuli that were being smelled, which were pre-

sented in opaque jars covered with permeable gauze coverings. Immediately after donation and prior to their use as stimuli, the tampons had been stored at $-60°$ C. As with the case of axillary odors, an inverse relationship was present between the intensity and pleasantness of the stimuli. In general, the vaginal odors were rated as less intense and less unpleasant mid-cycle, but, on average, were not perceived as pleasant. The individual pleasantness (or, perhaps more appropriately, unpleasantness) estimates were quite variable, suggesting that on an individual basis such odors were poor determinants of time of ovulation. Nonetheless, given the small number of donors and the fact that, in some cases, the odors were much more neutral mid-cycle, such secretions may influence sexual behavior in parallel with cultural taboos against intercourse during the period of menstrual bleeding. Unlike axillary secretions, there have been very few studies examining the effects of vaginal odors on the psychology or physiology of humans. Like axillary secretions and attempts to find vaginal pheromones in other animals, no human vaginal pheromone has been identified, the influence of learning on such responses is unknown, and research in this area has dwindled.

2. Bhatnagar and Meisami (1998) have argued that three components of a "vomeronasal organ complex" are needed to define a functioning vomeronasal system; namely, the tubular vomeronasal organ proper with appropriate sensory elements, the vomeronasal nerve, and an accessory olfactory bulb connected to the medial amygdala, implying the human VNO is nonfunctional. They point out that one, two, or all three of these elements can be present in members of families within the same mammalian order. This is the case in the 18 families of bats, where variable expression of the 3 components is observed. With rare exception, all of the necessary elements are found in only one family of bats, the New World leaf-nosed bats (Phyllostomidae). An extended set of criteria are listed by Evans (2006) for a functional accessory olfactory system: a patent duct system that transports stimuli to the VNO, a cartilage that typically surrounds the VNO, a multilayered VNO neuroepithelium, a vasomotor system, a sensory tract made up of VNO axons and their primary first-order synapses within epithelial receptive fields, and the accessory olfactory bulb and its subcortical nuclei.

References

....

Abel, E.L. (1991) *Alarm substance emitted by rats in the forced-swim test is a low volatile pheromone*. Physiol. Behav., 50, 723–727.

Adams, C. (2002) *Does menstrual synchrony really exist? The Straight Dope. The Chicago Reader*, Dec. 20.

Alberts, J.R. and Brunjes, P.C. (1978) *Ontogeny of thermal and olfactory determinants of huddling in the rat*. J. Comp. Physiol. Psychol., 92, 897–906.

Alberts, J.R. and Galef, B.G. (1973) *Olfactory cues and movement: stimuli mediating intraspecific aggression in the wild Norway rat*. J. Comp. Physiol. Psychol., 85, 233–242.

Alberts, J.R. and May, B. (1984) *Nonnutritive, thermotactile induction of filial huddling in rat pups*. Dev. Psychobiol., 17, 161–181.

Albone, E.S. (1984) Mammalian semiochemistry. John Wiley & Sons, New York.

Albonetti, M.E. and D'udine, B. (1986) *Social experience occurring during adult life: its effects on socio-sexual olfactory preferences in inbred mice*, Mus musculus. Anim. Behav., 34, 1844–1847.

Alexander, G. and Stevens, D. (1985) *Fostering in sheep. III. Facilitation by the use of odorants*. Appl. Anim. Beh. Sci., 14, 335–344.

Allin, J.T. and Banks, E.M. (1972) *Functional aspects of ultra-sound reproduction by infant Albino rats* (Rattus norvegicus). Anim. Behav., 20, 175–185.

Ameli, M. and De Marini, M. (1966) *Carateristiche morfologiche ed istiochimiche dell'endometrio di ratto a seguito di stimolazione acustica prolongata [Morphological and histochemical characteristics of the rat endometrium after prolonged acoustic stimulation]*. Clin. Otorinolarin., 18, 354–379.

Amoore, J.E., Pelosi, P. and Forrester, L.J. (1977) Specific anosmias to 5-α-androst-16-en-3-one and omega-pentadecalactone—ruinous and musky primary odors. *Chem. Senses Flav.* 2, 401–425.

Andervont, H.B. (1944) *Influence of environment on mammary cancer in mice*. J. Natl. Cancer Inst., 4, 579–581.

Anonymous. (1970) *Effects of sexual activity on beard growth in man*. Nature, 226, 869–870.

Anonymous. (2005) *So much more to know*. Science, 309, 78–102.

Arai, K., Kuwabara, Y. and Okinaga, S. (1972) *The effect of adrenocorticotropic hormone and dexamethasone, administered to the fetus in utero, upon maternal and fetal estrogens*. Amer. J. Obstet. Gyncol., 113, 316–322.

Arden, M.A. and Dye, L. (1998) *The assessment of menstrual synchrony: comment on Weller and Weller (1997)*. J. Comp. Psychol., 112, 323–324.

Arnould, C., Rousmans, S. and Vernet-Maury, E. (1994) *Influence of dodecyl propionate, attractive pheromone from rat pups, on rats' food intake*. Adv. Biosci., 93, 377–382.

Aron, C. (1979) *Mechanisms of control of the reproductive function by olfactory stimuli in female mammals*. Physiol. Rev., 59, 229–284.

Árvay, A. (1964) *Cortico-hypothalamic control of gonadotrophic functions*. In Bajusz, E. and Jasmin, G. (eds.), Major problems in neuroendocrinology. S. Karger, Basel, pp. 307–321.

Árvay, A. and Jasmin, G. (1964) *Cortico-hypothalamic control of gonadotrophic function*. In Bajusz, E. and Jasmin, G. (eds.), Major problems in neuroendocrinology. S. Karger, New York, pp. 307–321.

Árvay, A. and Nagy, T. (1959) *Der Einfluß belastender Nervenreize auf den Zeitpunkt des Eintrittes der Geschlechtsreife*. Acta Neurovegica, 20, 76.

Bakker, J. (2003) *Sexual differentiation of the neuroendocrine mechanisms regulating mate recognition in mammals*. J. Neuroendocrinol., 15, 615–621.

Balogh, R.D. and Porter, R.H. (1986) *Olfactory preferences resulting from mere exposure in human neonates*. Infant Beh. Dev., 9, 395–401.

Baran, D. (1973) *Responses of male Mongolian gerbils to male gerbil odors*. J. Comp. Physiol. Psychol., 84, 63–72.

Barfield, R.J., Auerbach, P., Geyer, L.A. and McIntosh, T.K. (1979) *Ultrasonic vocalizations in rat sexual behavior*. Amer. Zoologist, 19, 469–480.

Bartke, A. and Wolff, G.L. (1966) *Influence of lethal yellow (Ay) gene on estrous synchrony in mice*. Science, 153, 79.

Bartoshuk, L.M. and Beauchamp, G.K. (1994) *Chemical senses*. Ann. Rev. Psychol., 45, 419–449.

Batsell, W.R., Jr., Caperton, J. and Paschall, G. (1999) *Olfactory transmission of aversive information in rats*. Psychol. Rec., 49, 459–474.

Batty, J. (1978) *Acute changes in plasma testosterone levels and their relation to measures of sexual behaviour in the male house mouse* (Mus musculus). Anim. Behav., 26, 349–357.

Beach, F.A. (1949) *Response of male dogs to urine from females in heat*. J. Mammal., 30, 391–392.

Beach, F.A. (1950) *The snark was a boojum.* Amer. Psychol., 5, 115–124.

Beach, F.A. (1955) *The descent of instinct.* Psychol. Rev., 62, 401–410.

Beach, F.A. (1974) *Effects of gonadal hormones on urinary behavior in dogs.* Physiol. Behav., 12, 1005–1013.

Beach, F.A. and Jaynes, J. (1954) *Effects of early experience upon the behavior of animals.* Psychol. Bull., 51, 239–263.

Beach, F.A. and LeBoeuf, B.J. (1967) *Coital behavior in dogs. 1. Preferential mating in the bitch.* Anim. Behav., 15, 546–558.

Bean, N.J. (1982) *Olfactory and vomeronasal mediation of ultrasonic vocalizations in male mice.* Physiol. Behav., 28, 31–37.

Bear, M., Connors, B.W. and Paradiso, M.A. (2006) Neuroscience: exploring the brain. Lippincott, Williams & Wilkins, Baltimore.

Beardwood, C.J. (1982) *Hormonal changes associated with auditory stimulation.* In Beaumont, P.J.V. and Burrows, G.D. (eds.), Handbook of psychiatry and endocrinology. Elsevier, Amsterdam, pp. 401–439.

Beauchamp, G.K. (1973) *Attraction of male guinea pigs to conspecific urine.* Physiol. Behav., 10, 589–594.

Beauchamp, G.K. (1974) *The perineal scent gland and social dominance in the male guinea pig.* Physiol. Behav., 13, 669–673.

Beauchamp, G.K. (1976) *Diet influences attractiveness of urine in guinea pigs.* Nature, 263, 587–588.

Beauchamp, G.K. (2000) *Defining pheromones.* In Stein, L.J. (ed.), The Monell connection. Monell Chemical Senses Center, Philadelphia, p. 2.

Beauchamp, G.K., Doty, R.L., Moulton, D.G. and Mugford, R.A. (1976) *The pheromone concept in mammals: a critique.* In Doty, R.L. (ed.), Mammalian olfaction, reproductive processes, and behavior. Academic Press, New York, pp. 143–160.

Beauchamp, G.K., Doty, R.L., Moulton, D.G. and Mugford, R.A. (1979) *In defense of the term pheromone: response by Beauchamp et al.* J. Chem. Ecol., 5, 301–305.

Beauchamp, G.K., Martin, I.G., Wysocki, C.J. and Wellington, J.L. (1982) *Chemoinvestigatory and sexual behavior of male guinea pigs following vomeronasal organ removal.* Physiol. Behav., 29, 329–336.

Beauchamp, G.K. and Wellington, J.L. (1984) *Habituation to individual odors occurs following brief, widely-spaced presentations.* Physiol. Behav., 32, 511–514.

Beauchamp, G.K., Yamazaki, K. and Boyse, E.A. (1985) *The chemosensory recognition of genetic individuality.* Sci. Amer., 253, 86–92.

Beaumont, P.J.V. (1982) Handbook of psychiatry and endocrinology. Elsevier, Amsterdam.

Beck, M. and Galef, B.G., Jr. (1989) *Social influences on the selection of a protein-sufficient diet by Norway rats* (Rattus norvegicus). J. Comp. Psychol., 103, 132–139.

Bellringer, J.F., Pratt, H.P. and Keverne, E.B. (1980) *Involvement of the vomeronasal organ and prolactin in pheromonal induction of delayed implantation in mice.* J. Reprod. Fertil., 59, 223–228.

Belluscio, L., Koentges, G., Axel, R. and Dulac, C. (1999) *A map of pheromone receptor activation in the mammalian brain.* Cell, 97, 209–220.

Bensafi, M., Tsutsui, T., Khan, R., Levenson, R.W. and Sobel, N. (2004) *Sniffing a human sex-steroid derived compound affects mood and autonomic arousal in a dose-dependent manner.* Psychoneuroendocrinology, 29, 1290–1299.

Benton, D. (1982) *The influence of androstenol—a putative human pheromone—on mood throughout the menstrual cycle.* Biol. Psychol., 15, 249–256.

Berglund, H., Lindstrom, P., Dhejne-Helmy, C. and Savic, I. (2008) *Male-to-female transsexuals show sex-atypical hypothalamus activation when smelling odorous steroids.* Cereb. Cortex., 18, 1900–1908.

Berglund, H., Lindstrom, P. and Savic, I. (2006) *Brain response to putative pheromones in lesbian women.* Proc. Nat. Acad. Sci. USA, 103, 8269–8274.

Berliner, D.L., Monti-Bloch, L., Jennings-White, C. and Diaz-Sanchez, V. (1996) *The functionality of the human vomeronasal organ (VNO): evidence for steroid receptors.* J. Steroid Biochem. Mol. Biol., 58, 259–265.

Bernhardt, P.C., Dabbs, J.M.J., Fielden, J.A. and Lutter, C.D. (1998) *Testosterone changes during vicarious experiences of winning and losing among fans at sporting events.* Physiol. Behav., 65, 59–62.

Bethe, A. (1932) *Vernachlässigte Hormone.* Naturwissenschaften, 11, 177–181.

Beynon, R.J. and Hurst, J.L. (2004) *Urinary proteins and the modulation of chemical scents in mice and rats.* Peptides, 25, 1553–1563.

Bhatnagar, K.P. and Meisami, E. (1998) *Vomeronasal organ in bats and primates: extremes of structural variability and its phylogenetic implications.* Microsci. Res. Tech., 43, 465–475.

Bhatnagar, K.P., Smith, T.D. and Winstead, W. (2002) *The human vomeronasal organ: part IV. Incidence, topography, endoscopy, and ultrastructure of the nasopalatine recess, nasopalatine fossa, and vomeronasal organ.* Amer. J. Rhinol., 16, 343–350.

Bilko, A., Altbacker, V. and Hudson, R. (1994) *Transmission of food preference in the rabbit: the means of information transfer.* Physiol. Behav., 56, 907–912.

Bird, S. and Gower, D.B. (1981) *The validation and use of a radioimmunoassay for 5 α-androst-16-en-3-one in human axillary collections.* J. Steroid Biochem., 14, 213–219.

Bitterman, M.E. (1965) *The evolution of intelligence*. Sci. Amer., 212, 92–100.

Black, S.L. (2001) *Does smelling granny relieve depressive mood? Commentary on "Rapid mood change and human odors."* Biol. Psychol., 55, 215–225.

Black, S.L. and Biron, C. (1982) *Androstenol as a human pheromone: no effect on perceived physical attractiveness*. Behav. Neur. Biol., 34, 326–330.

Bloch, S. (1971) *Enhancement of on-time nidations in suckling pregnant mice by the proximity of strange males*. J. Endocrinol., 49, 431–436.

Bloch, S. (1974) *Observations on the ability of the stud male to block pregnancy in the mouse*. J. Reprod. Fertil., 38, 469–471.

Bloch, S. (1976) *Ein progstgeron abhangiges pheromon der weiblichen Maus [A progesterone-dependent pheromone of the female mouse]*. Experientia, 32, 937–938.

Bloch, S. and Wyss, H.I. (1973) *An anti-androgen (cyproterone acetate) inhibits the pregnancy block in mice caused by the presence of strange males (Bruce effect)*. J. Endocrinol., 59, 365–366.

Blum, S.L., Balsiger, D., Ricci, J.S. and Spiegel, D.K. (1975) *Effects of early exposure to ventral gland odor on physical and behavioral development and adult social behavior in Mongolian gerbils*. J. Comp. Physiol. Psychol., 89, 1210–1219.

Boissy, A., Terlouw, C. and Le Neindre, P. (1998) *Presence of cues from stressed conspecifics increases reactivity to aversive events in cattle: evidence for the existence of alarm substances in urine*. Physiol. Behav., 63, 489–495.

Bond, A. and Lader, M. (1974) *Use of analog scales in rating subjective feelings*. Brit. J. Med. Psychol., 47, 211–218.

Bonsall, R.W. and Michael, R.P. (1971) *Volatile constituents of primate vaginal secretions*. J. Reprod. Fertil., 27, 478–479.

Booth, A., Shelley, G., Mazur, A., Tharp, G. and Kittok, R. (1989) *Testosterone, and winning and losing in human competition*. Horm. Behav., 23, 556–571.

Booth, K.K. and Katz, L.S. (2000) *Role of the vomeronasal organ in neonatal offspring recognition in sheep*. Biol. Reprod., 63, 953–958.

Booth, W.D. (1980) *Endocrine and exocrine factors in the reproductive behaviour of the pig*. Symp. Zool. Soc. Lond., 45, 289–311.

Boulkroune, N., Wang, L.W., March, A., Walker, N. and Jacob, T.J.C. (2007) *Repetitive olfactory exposure to the biologically significant steroid androstadienone causes a hedonic shift and gender dimorphic changes in olfactory-evoked potentials*. Neuropsychopharmacology, 32, 1822–1829.

Boyar, R.M., Finkelstein, J.W., David, R., Roffwarg, H., Kapen, S., Weitzman, E.D. and Hellman, L. (1973) *Twenty-four hour patterns of plasma luteinizing hormone and follicle-stimulating hormone in sexual precocity*. N. Engl. J. Med., 289, 282–286.

Boyle, G.J. (1984) *Reliability and Validity of Izards Differential Emotions Scale.* Pers. Indiv. Diff., 5, 747–750.

Boyle, J.A., Lundstrom, J.N., Knecht, M., Jones-Gotman, M., Schaal, B. and Hummel, T. (2006) *On the trigeminal percept of androstenone and its implications on the rate of specific anosmia.* J. Neurobiol., 66, 1501–1510.

Brain, P.F., Homady, M.H., Castano, D. and Parmigiani, S. (1987) *Pheromones and behaviour of rodents and primates.* Boll. Zool., 4, 279–288.

Brain, P.F. and Nowell, N.W. (1971) *Isolation versus grouping effects on adrenal and gonadal function in albino mice. II. The female.* Gen. Comp. Endocrinol., 16, 155–159.

Brechbühl, J., Klaey, M. and Broillet, M.-C. (2008) *Grueneberg ganglion cells mediate alarm pheromone detection in mice.* Science, 321, 1092–1095.

Breen, M.F. and Leshner, A.I. (1977) *Maternal pheromone: a demonstration of its existence in the mouse* (Mus musculus). Physiol. Behav., 18, 527–529.

Brennan, P., Kaba, H. and Keverne, E.B. (1990) *Olfactory recognition: a simple memory system.* Science, 250, 1223–1226.

Brennan, P.A. and Keverne, E.B. (1997) *Neural mechanisms of mammalian olfactory learning.* Prog. Neurobiol., 51, 457–481.

Brennan, P.A. and Keverne, E.B. (2004) *Something in the air? New insights into mammalian pheromones.* Curr. Biol., 14, R81–R89.

Briand, L., Huet, J., Perez, V., Lenoir, G., Nespoulous, C., Boucher, Y., Trotier, D. and Pernollet, J.C. (2000) *Odorant and pheromone binding by aphrodisin, a hamster aphrodisiac protein.* FEBS Lett., 476, 179–185.

Bronson, F.H. (1968) *Pheromonal influences on mammalian reproduction.* In Diamond, M. (ed.), Pheromonal influences on mammalian reproduction. Indiana University Press, Bloomington, pp. 341–361.

Bronson, F.H. (1976b) *Serum FSH, LH, and prolactin in adult ovariectomized mice bearing silastic implants of estradiol: responses to social cues.* Biol. Reprod., 15, 147–152.

Bronson, F.H. (1976a) *Urine marking in mice: causes and effects.* In Doty, R.L. (ed.), Mammalian olfaction, reproductive processes and behavior. Academic Press, New York, pp. 119–143.

Bronson, F.H. (1979) *The reproductive ecology of the house mouse.* Quart. Rev. Biol., 54, 265–299.

Bronson, F.H. and Coquelin, A. (1980) *The modulation of reproduction by priming pheromones in housemice: speculations on adaptive function.* In Müller-Schwarze, D. and Silverstein, R.M. (eds.), Chemical signals in vertebrates. Plenum Press, New York, pp. 243–265.

Bronson, F.H. and Desjardins, C. (1969) *Release of gonadotrophin in ovariectomized mice after exposure to males.* J. Endocrinol., 44, 293–297.

Bronson, F.H. and Eleftheriou, B.E. (1963) *Influence of strange males on implantation in the deer mouse.* Gen. Comp. Endocrinol., 3, 515–518.

Bronson, F.H., Eleftheriou, B.E. and Dezell, H.E. (1969) *Strange male pregnancy block in deermice: prolactin and adrenocortical hormones.* Biol. Reprod., 1, 302–306.

Bronson, F.H. and Maruniak, J.A. (1975) *Male-induced puberty in female mice: evidence for a synergistic action of social cues.* Biol. Reprod., 13, 94–98.

Brooksbank, B.W.L., Brown, R. and Gustafsson, J.A. (1974) *The detection of 5α-androst-16-en-3α-ol in human male axillary sweat.* Experientia, 30, 864–865.

Brooksbank, B.W.L. and Haslewood, G.A.D. (1961) *The estimation of androsten-16-en-3α-ol in human urine.* Biochem. J., 80, 488–496.

Brouette-Lahlou, I., Vernet-Maury, E. and Vigouroux, M. (1992) *Role of pups' ultrasonic calls in a particular maternal behavior in Wistar rat: pups' anogenital licking.* Behav. Brain Res., 50, 147–154.

Brown, G.M., Seggie, J. and Feldmann, J. (1977) *Effect of psychosocial stimuli and limbic lesions on prolactin at rest and following stress.* Clin. Endocrinol., 6 Suppl., 29S–41S.

Brown, R.E. (1977) *Odor preference and urine-marking scales in male and female rats: effects of gonadectomy and sexual experience on responses to conspecific odors.* J. Comp. Physiol. Psychol., 91, 1090–1206.

Brown, R.E. (1979) *Mammalian social odors: a critical review.* Advances in the Study of Behavior. Academic Press, New York, pp. 103–162.

Brown, R.E. (1985) *Introduction: the pheromone concept in mammalian chemical communication.* In Brown, R.E. and MacDonald, D.W. (eds.), Social odours in mammals. Clarendon Press, Oxford, pp. 1–18.

Brown, R.E. (1988) *Individual odors of rats are discriminable independently of changes in gonadal hormone levels.* Physiol. Behav., 43, 359–363.

Brown, R.E. and MacDonald, D.W. (1985) Social odours in mammals. Clarendon Press, Oxford.

Brown, S.M. and Lisk, R.D. (1978) *Blocked sexual receptivity in grouped female golden hamsters, the result of contact induced inhibition.* Biol. Reprod., 18, 829–833.

Brown, W.L., Jr. (1968) *An hypothesis concerning the function of the metapleural glands in ants.* Amer. Naturalist, 102, 188–191.

Brownlee, R.G., Silverstein, R.M., Müller-Schwarze, D. and Singer, A.G. (1969) *Isolation, identification, and function of the chief component of the male tarsal scent in black-tailed deer.* Nature, 221, 284–285.

Bruce, H.M. (1959) *An exteroceptive block to pregnancy in the mouse.* Nature, 184, 105.

Bruce, H.M. (1960a) *A block to pregnancy in the mouse caused by the proximity of strange males.* J. Reprod. Fertil., 1, 96–103.

Bruce, H.M. (1960b) *Further observations on pregnancy block in mice caused by the proximity of strange males.* J. Reprod. Fertil., 1, 311–312.

Bruce, H.M. (1961) *Time relations in the pregnancy-block induced in mice by strange males.* J. Reprod. Fertil., 2, 138–142.

Bruce, H.M. (1963) *Olfactory block to pregnancy among grouped mice.* J. Reprod. Fertil., 6, 451.

Bruce, H.M. (1965) *Effect of castration on the reproductive pheromones of male mice.* J. Reprod. Fertil., 10, 141–143.

Bruce, H.M. (1968) *Absence of pregnancy-block in mice when stud and test males belong to an inbred strain.* J. Reprod. Fertil., 17, 407–408.

Bruce, H.M. (1970) *Pheromones.* Brit. Med. Bull., 26, 10–13.

Bruce, H.M., Land, R.B. and Falconer, D.S. (1968) *Inhibition of pregnancy-block in mice by handling.* J. Reprod. Fertil., 15, 289–294.

Bruce, H.M. and Parkes, A.S. (1960) *Hormonal factors in exteroreceptive block to pregnancy in mice.* J. Endocrinol., 20, 29–30.

Bruce, H.M. and Parkes, A.S. (1961) *The effect of concurrent lactation on the olfactory block to pregnancy in the mouse.* J. Reprod. Fertil., 6, vi–vii.

Bruce, H.M. and Parrott, D.M.V. (1960) *Role of olfactory sense in pregnancy block by strange males.* Science, 131, 1526.

Brunjes, P.C. (1992) *Lessons from lesions—the effects of olfactory bulbectomy.* Chem. Senses, 17, 729–763.

Buck, L.B. (2000) *The molecular architecture of odor and pheromone sensing in mammals.* Cell, 100, 611–618.

Burgdorf, J., Knutson, B. and Panksepp, J. (2000) *Anticipation of rewarding electrical brain stimulation evokes ultrasonic vocalization in rats.* Behav. Neurosci., 114, 320–327.

Butenandt, A., Beckmann, R., Stamm, D. and Hecker, E. (1959) *Uber den Sexual-Lockstoff des Seidenspinners Bombyx Mori—Reindarstellung und Konstitution.* Z. Naturfo. B., 14, 283–284.

Byatt, S. and Nyby, J. (1986) *Hormonal regulation of chemosignals of female mice that elicit ultrasonic vocalizations from males.* Horm. Behav., 20, 60–72.

Campbell, N.A. (1996) Biology. Benjamin/Cummings Publishing Co., Inc., Menlo Park, CA.

Carani, C., Bancroft, J., Del, R.G., Granata, A.R., Facchinetti, F. and Marrama, P. (1990) *The endocrine effects of visual erotic stimuli in normal men.* Psychoneuroendocrinology, 15, 207–216.

Caroom, D. and Bronson, F.H. (1971) *Responsiveness of female mice to preputial attractant: effects of sexual experience and ovarian hormones.* Physiol. Behav., 7, 659–662.

Carr, W.J. and Caul, W.F. (1962) *The effect of castration in rat upon the discrimination of sex odours*. Anim. Behav., 10, 20–27.

Carr, W.J., Loeb, L.S. and Dissinger, M.E. (1965) *Responses of rats to sex odors*. J. Comp. Physiol. Psychol., 59, 370–377.

Carr, W.J., Loeb, L.S. and Wylie, N.R. (1966) *Responses to feminine odors in normal and castrated male rats*. J. Comp. Physiol. Psychol., 62, 336–338.

Carr, W.J., Solberg, B. and Pfaffman, C. (1962) *The olfactory threshold for estrous female urine in normal and castrated male rats*. J. Comp. Physiol. Psychol., 55, 415–417.

Carr, W.J., Yee, L., Gable, D. and Marasco, E. (1976) *Olfactory recognition of conspecifics by domestic Norway rats*. J. Comp. Physiol. Psychol., 90, 821–828.

Carroll, L. (1874) The hunting of the snark: an agony in eight fits. Macmillan, New York.

Carter, C.S. (1972) *Effects of olfactory experience on the behaviour of the guinea-pig* (Cavia porcellus). Anim. Behav., 20, 54–60.

Carter, C.S. and Marr, J.N. (1970) *Olfactory imprinting and age variables in the guinea-pig*, Cavia porcellus. Anim. Behav., 18, 238–244.

Castro, B.M. (1967) *Age of puberty in female mice: relationship to population density and the presence of adult males*. Acad. Brasil. Cienc., 39, 289–291.

Cepicky, P., Mandys, F., Hlavicka, L. and Sosnova, K. (1996) *Absence of menstrual cycle synchronization in mentally affected women living in a social welfare institute*. Homeostasis Health Dis., 37, 249–252.

Chamero, P., Marton, T.F., Logan, D.W., Flanagan, K., Cruz, J.R., Saghatelian, A., Cravatt, B.F. and Stowers, L. (2007) *Identification of protein pheromones that promote aggressive behaviour*. Nature, 450, 899–902.

Champlin, A.K. (1971) *Suppression of oestrus in grouped mice: the effects of various densities and the possible nature of the stimulus*. J. Reprod. Fertil., 27, 233–241.

Chapman, V.M., Desjardins, C. and Whitten, W.K. (1970) *Pregnancy block in mice: changes in pituitary LH and LTH and plasma progestin levels*. J. Reprod. Fertil., 21, 333–337.

Cheetham, S.A., Thom, M.D., Jury, F., Ollier, W.E.R., Beynon, R.J. and Hurst, J.L. (2007) *The genetic basis of individual-recognition signals in the mouse*. Curr. Biol., 17, 1771–1777.

Chen, D. and Haviland-Jones, J. (1999) *Rapid mood change and human odors*. Physiol. Behav., 68, 241–250.

Chipman, R.K. and Albrecht, E.D. (1974) *The relationship of the male preputial gland to the acceleration of oestrus in the laboratory mouse*. J. Reprod. Fertil., 38, 91–96.

Chipman, R.K. and Fox, K.A. (1966a) *Oestrous synchronization and pregnancy blocking in wild house mice* (Mus musculus). J. Reprod. Fertil., 12, 233–236.

Chipman, R.K. and Fox, K.A. (1966b) *Factors in pregnancy blocking: age and reproductive background of females: numbers of strange males.* J. Reprod. Fertil., 12, 399–403.

Chipman, R.K., Holt, J.A. and Fox, K.A. (1966) *Pregnancy failure in laboratory mice after multiple short-term exposure to strange males.* Nature, 210, 653.

Christian, J.J. (1956) *Reserpine suppression of density-dependent adrenal hypertrophy and reproductive hypoendocrinism in populations of male mice.* Amer. J. Physiol., 187, 356.

Christian, J.J. (1959) *Control of population growth in rodents by interplay between population density and endocrine physiology.* Wildl. Dis., 2, 1–32.

Christian, J.J. (1960) *Adrenocortical and gonadal responses of female mice to increased population density.* Proc. Soc. Expl. Biol. Med., 104, 330–332.

Christian, J.J. and Davis, D.E. (1964) *Endocrines, behavior, and population.* Science, 146, 1550–1560.

Christian, J.J. and LeMunyan, C.D. (1958) *Adverse effects on crowding on reproduction and lactation of mice and two generations of their progeny.* Endocrinology, 63, 517–529.

Christian, J.J., Lloyd, J.A. and Davis, D.E. (1965) *The role of endocrines in the self-regulation of mammalian populations.* Rec. Prog. Horm. Res., 21, 501–568.

Clancy, A.N., Macrides, F., Singer, A.G. and Agosta, W.C. (1984) *Male hamster copulatory responses to a high molecular weight fraction of vaginal discharge: effects of vomeronasal organ removal.* Physiol. Behav., 33, 653–660.

Clark, A.B. (1982) *Scent marks as social signals in* Galago crassicaudatus. *II. Discrimination between individuals by scent.* J. Chem. Ecol., 8, 1153–1165.

Clark, A.J., Charal, P., Bingham, R.W., Barrett, D. and Bishop, J.O. (1985) *Sequence structures of a mouse major urinary protein gene and pseudogene compared.* EMBO Journal, 4, 3159–3165.

Clark, M.M. and Galef, B.J. Jr. (2002) *Socially induced delayed reproduction in female Mongolian gerbils* (Meriones unguiculatus): *is there anything special about dominant females?* J. Comp. Psychol., 116, 363–368.

Claus, R. and Hoppen, H.O. (1979) *The boar-pheromone steroid identified in vegetables.* Experientia, 35, 1674–1675.

Clee, M.D., Humphreys, E.M. and Russell, J.A. (1975) *The suppression of ovarian cyclical activity in groups of mice, and its dependence on ovarian hormones.* J. Reprod. Fertil., 45, 395–398.

Clegg, F. and Williams, D. (1983) *Maternal pheromone in* Rattus norvegicus. Behav. Neur. Biol., 37, 223–236.

Clulow, F.V. and Clarke, J.E. (1968) *Pregnancy-block in* Microtus agrestis, *an induced ovulator.* Nature, 219, 511.

Clulow, F.V. and Langford, P.E. (1964) *Pregnancy-block in the meadow vole* Microtus pennsylvanicus. J. Reprod. Fertil., 24, 275–277.

Cohen-Tannoudji, J., Einhorn, J. and Signoret, J.P. (1994) *Ram sexual pheromone: first approach of chemical identification.* Physiol. Behav., 56, 955–961.

Cohen-Tannoudji, J., Locatelli, A. and Signoret, J.P. (1986) *Non-pheromonal stimulation by the male of LH release in the anoestrous ewe.* Physiol. Behav., 36, 921–924.

Colby, D.R. and Vandenberg, J.G. (1974) *Regulatory effects of urinary pheromones on puberty in the mouse.* Biol. Reprod., 11, 268–279.

Comfort, A. (1971) *Likelihood of human pheromones.* Nature, 230, 432–433.

Cooper, K.J. and Haynes, N.B. (1967) *Modification of the oestrous cycle of the under-fed rat associated with the presence of the male.* J. Reprod. Fertil., 14, 317–320.

Cooper, K.J., Purvis, K. and Haynes, N.B. (1972) *Further observations on the ability of the male to influence the oestrous cycle of the underfed rat.* J. Reprod. Fertil., 28, 473–475.

Coopersmith, R. and Leon, M. (1984) *Enhanced neural response to familiar olfactory cues.* Science, 225, 849–851.

Coquelin, A., Clancy, A.N., Macrides, F., Noble, E.P. and Gorski, R.A. (1984) *Pheromonally induced release of luteinizing hormone in male mice: involvement of the vomeronasal system.* J. Neurosci., 4, 2230–2236.

Coria-Avila, G.A., Ouimet, A.J., Pacheco, P., Manzo, J. and Pfaus, J.G. (2005). *Olfactory conditioned partner preference in the femal rat.* Behav. Neurosci., 119, 716–725.

Cornwell, C.A. (1976) *Selective olfactory exposure alters social and plant odor preferences of immature hamsters.* Behav. Biol., 17, 131–137.

Coureaud, G., Langlois, D., Sicard, G. and Schaal, B. (2004) *Newborn rabbit responsiveness to the mammary pheromone is concentration-dependent.* Chem. Senses, 29, 341–350.

Coureaud, G., Schaal, B., Coudert, P., Rideaud, P., Fortun-Lamothe, L., Hudson, R. and Orgeur, P. (2000) *Immediate postnatal sucking in the rabbit: its influence on pup survival and growth.* Reprod. Nutr. Develop., 40, 19–32.

Coureaud, G., Schaal, B., Langlois, D. and Perrier, G. (2001) *Orientation response of newborn rabbits to odours of lactating females: relative effectiveness of surface and milk cues.* Anim. Behav., 61, 153–162.

Coureaud, G., Schaal, B., Orgeur, P., Hudson, R., Lebas, F. and Coudert, P. (1997) *Perinatal odour disruption impairs neonatal milk intake in the rabbit.* Adv. Ethol., 32, 134.

Cousens, G. and Otto, T. (1998) *Both pre- and posttraining excitotoxic lesions of the basolateral amygdala abolish the expression of olfactory and contextual fear conditioning.* Behav. Neurosci., 112, 1092–1103.

Cowley, J.J. and Brooksbank, B.W. (1991) *Human exposure to putative pheromones and changes in aspects of social behaviour.* J. Steroid Biochem. Mol. Biol., 39, 647–659.

Cowley, J.J., Johnson, A.L. and Brooksbank, B.W. (1977) *The effect of two odorous compounds on performance in an assessment-of-people test.* Psychoneuroendocrinology, 2, 159–172.

Cowley, J.J. and Pewtress, R.K. (1986) *Ano-genital distance as a factor in determining puberty acceleration in mice.* J. Reprod. Fertil., 78, 685–691.

Cowley, J.J. and Wise, D.R. (1972) *Some effects of mouse urine on neonatal growth and reproduction.* Anim. Behav., 20, 499–506.

Craig, W. (1918) *Appetites and aversions as constituents of instincts.* Biol. Bull., 34, 91–107.

Crew, F.A. and Mirskaia, L. (1931) *The effects of density on an adult mouse population.* Biol. Generalis, 7, 239–250.

Crowcroft, P. and Rowe, F.P. (1958) *The growth of confined colonies of the wild house-mouse* (Mus musculus): *the effect of dispersal on female fecundity.* Proc. Zool. Soc. Lond., 131, 357–365.

Curtis, R.F., Ballantine, J.A., Keveren, E.B., Bonsall, R.W. and Michael, R.P. (1971) *Identification of primate sexual pheromones and the properties of synthetic attractants.* Nature, 232, 396–398.

Cutler, W.B., Friedmann, E. and McCoy, N.L. (1998) *Pheromonal influences on sociosexual behavior in men.* Arch. Sex. Behav., 27, 1–13.

Cutler, W.B. and Genovese, E. (2002) *Pheromones, sexual attractiveness and quality of life in menopausal women.* Climacteric, 5, 112–121.

Cutler, W.B. and Genovese-Stone, E. (1998) *Wellness in women after 40 years of age: the role of sex hormones and pheromones.* Disease-A-Month, 44, 421–546.

D'Amato, F.R. and Cabib, S. (1987) *Chronic exposure to a novel odor increases pups' vocalizations, maternal care, and alters dopaminergic functioning in developing mice.* Behav. Neur. Biol., 48, 197–205.

D'Amato, F.R. and Cabib, S. (1990) *Behavioral effects of manipulations of the olfactory environment in developing mice: involvement of the dopaminergic system.* In Puglisi-Allegra, S. and Oliverio, A. (eds.), Psychobiology of stress. Kluwer Academic Publishers, Amsterdam, pp. 59–71.

Dardes, R.C., Baracat, E.C. and Simoes, M.J. (2000) *Modulation of estrous cycle and LH, FSH and melatonin levels by pinealectomy and sham-pinealectomy in female rats.* Prog. Neuro-Psychopharm. Biol. Psychiat., 24, 441–453.

Dawkins, R. and Krebs, J.R. (1978) *Animal signals: information or manipulation?* In Krebs, J.R. and Davies, N.B. (eds.), Behavioural ecology: an evolutionary approach. Blackwood, Oxford, pp. 282–309.

deCatanzaro, D. (1988) *Effect of predator exposure upon early pregnancy in mice*. Physiol. Behav., 43, 691–696.

deCantanzaro, D., Baptista, M.A. and Spironello-Vella, E. (2001) *Administration of minute quantities of 17beta-estradiol on the nasal area terminates early pregnancy in inseminated female mice*. Pharmacol. Biochem. Behav., 69, 503–509.

deCatanzaro, D. and Graham, C. (1992) *Influences of exogenous epinephrine on two reproductive parameters in female mice: disruption of receptivity but not implantation*. Horm. Behav., 26, 330–338.

deCatanzaro, D. and MacNiven, E. (1992) *Psychogenic pregnancy disruptions in mammals*. Neurosci. Biobehav. Rev., 16, 43–53.

deCatanzaro, D., MacNiven, E., Goodison, T. and Richardson, D. (1994) *Estrogen antibodies reduce vulnerability to stress-induced failure of intrauterine implantation in inseminated mice*. Physiol. Behav., 55, 35–38.

deCatanzaro, D., MacNiven, E. and Ricciuti, F. (1991) *Comparison of the adverse effects of adrenal and ovarian steroids on early pregnancy in mice*. Psychoneuroendocrinology, 16, 525–536.

deCatanzaro, D., Muir, C., O'Brien, J. and Williams, S. (1995a) *Strange-male-induced pregnancy disruption in mice: reduction of variability by 17β-estradiol antibodies*. Physiol. Behav., 58, 401–404.

deCatanzaro, D., Muir, C., Sullivan, C. and Boissy, A. (1999) *Pheromones and novel male-induced pregnancy disruptions in mice: exposure to conspecifics is necessary for urine alone to induce an effect*. Physiol. Behav., 66, 153–157.

deCatanzaro, D. and Storey, A.E. (1989) *Partial mediation of strange-male-induced pregnancy blocks by sexual activity in mice* (Mus musculus). J. Comp. Psychol., 103, 381–388.

deCatanzaro, D., Wyngaarden, P., Griffiths, J., Ham, M., Hancox, J. and Brain, D. (1995b) *Interactions of contact, odor cues, and androgens in strange-male-induced early pregnancy disruptions in mice* (Mus musculus). J. Comp. Psychol., 109, 115–122.

Dember, W.N. and Jenkins, J.J. (1970) General psychology. Prentice-Hall, Inc., Englewood Cliffs, NJ.

Denenberg, V.H., Desantis, D., Waite, S. and Thoman, E.B. (1977) *The effects of handling in infancy on behavioral states in the rabbit*. Physiol. Behav., 18, 553–557.

Dewar, A.D. (1959) *Observations on pseudopregnancy in the mouse*. J. Endocrinol., 18, 186–190.

Distel, H. and Hudson, R. (1984) *Nipple-search pheromone in rabbits: dependence on season and reproductive state*. J. Comp. Physiol. A., 155, 13–17.

Distel, H. and Hudson, R. (1985) *The contribution of the olfactory and tactile modalities to the nipple-search behaviour of newborn rabbits*. J. Comp. Physiol. A, 157, 599–605.

Dixon, A.K. (1982) *A possible olfactory component in the effects of diazepam on social behavior of mice.* Psychopharmacologia, 77, 246–252.

Dizinno, G. and Whitney, G. (1977) *Androgen influences on male ultrasounds during courtship.* Horm. Behav., 8, 188–192.

Dizinno, G. and Whitney, G. (1978) *Ultrasonic vocalizations by male mice* (Mus musculus) *in response to female sex pheromone: experiential determinants.* Behav. Biol., 22, 104–113.

Dominguez, H.D., Lopez, M.F. and Molina, J.C. (1999) *Interactions between perinatal and neonatal associative learning defined by contiguous olfactory and tactile stimulation.* Neurobiol. Learn. Mem., 71, 272–288.

Dominic, C.J. (1964) *Source of the male odour causing pregnancy block in mice.* J. Reprod. Fertil., 8, 266–267.

Dominic, C.J. (1965) *The origin of the pheromones causing pregnancy block in mice.* J. Reprod. Fertil., 10, 469–472.

Dominic, C.J. (1966a) *Effects of single ectopic pituitary grafts on the oestrous cycle of the intact mouse.* J. Reprod. Fertil., 12, 533–538.

Dominic, C.J. (1966b) *Observations on the reproductive pheromones of mice. II. Neuro-endocrine mechanisms involved in the olfactory block to pregnancy.* J. Reprod. Fertil., 11, 415.

Dominic, C.J. (1966c) *Reserpine: inhibition of olfactory blockage of pregnancy in mice.* Science, 152, 1764–1765.

Dominic, C.J. (1967) *Effect of ectopic pituitary grafts on the olfactory block to pregnancy in mice.* Nature, 213, 1242.

Dorries, K.M., Adkins-Regan, E. and Halpern, B.P. (1997) *Sensitivity and behavioral responses to the pheromone androstenone are not mediated by the vomeronasal organ in domestic pigs.* Brain Behav. Evol., 49, 53–62.

Doty, R.L. (1972) *Odor preferences of female* Peromyscus maniculatus bairdi *for male mouse odors of* P. m. bairdi *and* P. leucopus noveboracensis *as a function of estrous state.* J. Comp. Physiol. Psychol., 81, 191–197.

Doty, R.L. (1973) *Reactions of deer mice* (Peromyscus maniculatus) *and white-footed mice* (Peromyscus leucopus) *to homospecific and heterospecific urine odors.* J. Comp. Physiol. Psychol., 84, 296–303.

Doty, R.L. (1974) *A cry for the liberation of the female rodent: courtship and copulation in Rodentia.* Psychol. Bull., 81, 159–172.

Doty, R.L. (1980) *Scent marking in mammals.* In Denny, M.R. (ed.), Comparative psychology: research in animal behavior. John Wiley & Sons, New York, pp. 385–399.

Doty, R.L. (1981) *Olfactory communication in humans.* Chem. Senses, 6, 351–376.

Doty, R.L. (1986a) *Gender and endocrine-related influences upon olfactory sen-*

sitivity. In Meiselman, H.L. and Rivlin, R.S. (eds.), Clinical measurement of taste and smell. Macmillan, New York, pp. 377–413.

Doty, R.L. (1986b) *Odor-guided behavior in mammals.* Experientia, 42, 257–271.

Doty, R.L. (2001) *Olfaction.* Ann. Rev. Psychol., 52, 423–452.

Doty, R.L. (2003a) *Cranial nerve I: olfactory nerve.* In Goetz, C.G. (ed.), Textbook of clinical neurology. W.B. Saunders, Philadelphia, pp. 99–110.

Doty, R.L. (2003b) Handbook of olfaction and gustation. 2nd ed. Marcel Dekker, New York.

Doty, R.L. (2008) *The olfactory vector hypothesis of neurodegenerative disease: Is it viable?* Ann. Neurol., 63, 7–15.

Doty, R.L. and Anisko, J.J. (1973) *Procaine hydrochloride olfactory block eliminates mounting in the male golden hamster.* Physiol. Behav., 10, 395–397.

Doty, R.L. and Cameron, L. (2009) *Sex differences and reproductive hormone influences on human odor perception,* Physiol. Behav., 97, 213–228.

Doty, R.L., Carter, C.S. and Clemens, L.G. (1971) *Olfactory control of sexual behavior in the male and early-androgenized female hamster.* Horm. Behav., 2, 325–335.

Doty, R.L. and Cometto-Muñiz, J.E. (2003) *Trigeminal chemosensation.* In Doty, R.L. (ed.), Handbook of olfaction and gustation. Marcel Dekker, New York, pp. 981–999.

Doty, R.L. and Dunbar, I. (1974) *Attraction of beagles to conspecific urine, vaginal and anal sac secretion odors.* Physiol. Behav., 12, 825–833.

Doty, R.L., Ford, M., Preti, G. and Huggins, G.R. (1975) *Changes in the intensity and pleasantness of human vaginal odors during the menstrual cycle.* Science, 190, 1316–1318.

Doty, R.L., Green, P.A., Ram, C. and Yankell, S.L. (1982a) *Communication of gender from human breath odors: relationship to perceived intensity and pleasantness.* Horm. Behav., 16, 13–22.

Doty, R.L., Hall, J.W., Flickinger, G.L. and Sondheimer, S.J. (1982b) *Cyclic changes in olfactory and auditory sensitivity during the menstrual cycle: no attenuation by oral contraceptive medication.* In Breipohl, W. (ed.), Olfaction and endocrine regulation. IRL Press, London, pp. 35–42.

Doty, R.L. and Kart, R. (1972) *A comparative and developmental analysis of the midventral sebaceous glands in 18 taxa of* Peromyscus, *with an examination of gonadal steroid influences in* Peromyscus maniculatus bairdii. J. Mammal., 53, 83–99.

Doty, R.L. and Laing, D.G. (2003) *Psychophysical measurement of olfactory function, including odorant mixture assessment.* In Doty, R.L. (ed.), Handbook of olfaction and gustation. Marcel Dekker, New York, pp. 203–228.

Doty, R.L., Orndorff, M.M., Leyden, J. and Kligman, A. (1978) *Communication of gender from human axillary odors: relationship to perceived intensity and hedonicity.* Behav. Neural. Biol., 23, 373–380.

Doty, R.L., Snyder, P.J., Huggins, G.R. and Lowry, L.D. (1981) *Endocrine, cardiovascular, and psychological correlates of olfactory sensitivity changes during the human menstrual cycle.* J. Comp. Physiol. Psychol., 95, 45–60.

Døving, K.B. and Trotier, D. (1998) *Structure and function of the vomeronasal organ.* J. Exp. Biol., 201, 2913–2925.

Drago, F., Amir, S., Continella, G., Alloro, M.C. and Scapagnini, U. (1986) *Effects of endogenous hyperprolactinemia on adaptive responses to stress.* In MacLeod, R.M., Thorner, M. and Scapagnini, U. (eds.), Prolactin—basic and clinical correlates. Liviana Press, Padova, pp. 609–614.

Drickamer, L.C. (1974) *Contact stimulation, androgenized females and accelerated sexual maturation in female mice.* Behav. Biol., 12, 101–110.

Drickamer, L.C. (1975) *Contact stimulation and accelerated sexual maturation of female mice.* Behav. Biol., 15, 113–115.

Drickamer, L.C. (1983) *Male acceleration of puberty in female mice* (Mus musculus). J. Comp. Psychol., 97, 191–200.

Drickamer, L.C. (1984a) *Acceleration of puberty in female mice by a chemosignal from pregnant and lactating females: circadian rhythm effects.* Biol. Reprod., 31, 104–108.

Drickamer, L.C. (1984b) *Acceleration of puberty in female mice by a urinary chemosignal from pregnant or lactating females: timing and duration of stimulation.* Dev. Psychobiol., 17, 451–455.

Drickamer, L.C. (1986a) *Puberty-influencing chemosignals in house mice: biological and evolutionary considerations.* In Duvall, D., Müller-Schwarze, D. and Silverstein, R.M. (eds.), Chemical signals in vertebrates 4. Plenum Press, New York, pp. 441–455.

Drickamer, L.C. (1986b) *Effects of urine from females in oestrus on puberty in female mice.* J. Reprod. Fertil., 77, 613–622.

Drickamer, L.C. (1987) *Intermittent stimulation and acceleration of puberty by urinary chemosignals in female mice* (Mus musculus domesticus). Biol. Reprod., 37, 89–95.

Dryden, G.L. and Conaway, C.H. (1967) *Origin and hormonal control of scent production in* Suncus murinus. J. Mammal., 48, 420–428.

Dukas, R. (2008) *Evolutionary biology of insect learning.* Ann. Rev. Entomol., 53, 145–160.

Dulac, C. and Axel, R. (1995) *A novel family of genes encoding putative pheromone receptors in mammals.* Cell, 83, 195–206.

Dulac, C. and Axel, R. (1998) *Expression of candidate pheromone receptor genes in vomeronasal neurons.* Chem. Senses, 23, 467–475.

Dunbar, I., Buehler, M. and Beach, F.A. (1980) *Developmental and activational effects of sex hormones on the attractiveness of dog urine.* Physiol. Behav., 24, 201–204.

Dunlap, K. (1919) *Are there any instincts?* J. Abn. Psychol., 14, 35–50.

Dunn, G.C., Price, E.O. and Katz, L.S. (1987) *Fostering calves by odor transfer.* Appl. Anim. Beh. Sci., 17, 33–39.

Ehrlichman, H. and Bastone, L. (1992) *The use of odour in the study of emotion.* In Toller, S. and Dodd, G.H. (eds.), Fragrance. The psychology and biology of perfume. Elsevier Applied Science, London, pp. 143–159.

Eibl-Eibesfeldt, I. (1970) Ethology: the biology of behavior. Holt, Rinehart and Winston, New York.

Einstein, A. (1934) *On the method of theoretical physics.* Phil. Sci., 1, 163–169.

Eleftheriou, B.E., Bronson, F.H. and Zarrow, M.X. (1962) *Interaction of olfactory and other environmental stimuli on implantation in the deer mouse.* Science, 137, 764.

Elias, M. (1981) *Serum cortisol, testosterone, and testosterone-binding globulin responses to competitive fighting in human males.* Aggr. Beh., 7, 215–224.

Ellis, B.J. and Garber, J. (2000) *Psychosocial antecedents of variation in girls' pubertal timing: Maternal depression, stepfather presence, and marital and family stress.* Child Development, 71, 485–501.

Endroczi, E. and Nyakas, C.S. (1974) *Pituitary-adrenal function in lactating rats.* Endokrinologie (Budapest), 63, 1–5.

Epley, S. (1974) *Reduction of the behavioral effects of aversive stimulation by the presence of companions.* Psychol. Bull., 81, 271–283.

Euker, J.S., Meites, J. and Riegle, G.D. (1975) *Effects of acute stress on serum LH and prolactin in intact, castrate and dexamethasone-treated male rats.* Endocrinology, 96, 85–92.

Euker, J.S. and Riegle, G.D. (1973) *Effects of stress on pregnancy in the rat.* J. Reprod. Fertil., 34, 343–346.

Evans, C. (2003) Vomeronasal chemoreception in vertebrates. Imperial College Press, London.

Evans, C.M. (1979) Studies on relationships between gonadal hormones and intra-specific aggression behaviour in rodents. University of Wales, Ph.D. thesis.

Evans, C.S. (2006) *Accessory chemosignaling mechanisms in primates.* Am. J. Primatol., 68, 525–544.

Ewer, R.F. (1968) Ethology of mammals. Elek Science, London.

Fenster, L., Waller, K., Chen, J., Hubbard, A.E., Windham, G.C., Elkin, E. and Swan, S. (1999) *Psychological stress in the workplace and menstrual function.* Am. J. Epidemiol., 149, 127–134.

Ferkin, M.H. (1999) *Scent over-marking and adjacent-marking as competitive* tactics used during chemical communication in voles. In Johnston, R.E., Müller-Schwarze, D. and Sorensen, P.W. (eds.), Advances in chemical signals in vertebrates. Kluwer Academic/Plenum Publishers, New York, pp. 239–246.

Filicori, M., Tabarelli, C., Casadio, P., Ferlini, F., Gessa, G., Pocognoli, P., Cognigni, G. and Pecorari, R. (1998) *Interaction between menstrual cyclicity and gonadotropin pulsatility.* Horm. Res., 49, 169–172.

Fillion, T.J. and Blass, E.M. (1986a) *Infantile behavioural reactivity to oestrous chemostimuli in Norway rats.* Anim. Behav., 34, 123–133.

Fillion, T.J. and Blass, E.M. (1986b) *Infantile experience with suckling odors determines adult sexual behavior in male rats.* Science, 231, 729–731.

Filsinger, E.E., Braun, J.J., Monte, W.C. and Linder, D.E. (1984) *Human* (Homo sapiens) *responses to the pig* (Sus scrofa) *sex pheromone 5 α-androst-16-en-3-one.* J. Comp. Psychol., 98, 219–222.

Flood, P. (1985) *Sources of significant smells: the skin and other organs.* In Brown, R.E. and MacDonald, D.W. (eds.), Social odours in mammals. Oxford University Press, Oxford, pp. 19–36.

French, J.A. and Stribley, J.A. (1985) *Patterns of urinary estrogen excretion in female golden lion tamarins (Leontopithecus-Rosalia).* J. Reprod. Fertil., 75, 537–546.

Fuchs, A.R., Cubile, L., Dawood, M.Y. and Jorgensen, F.N. (1984) *Release of oxytocin and prolactin by suckling rabbits throughout lactation.* Endocrinology, 114, 462–469.

Fullenkamp, A., Fischer, R.B., Vance, R.A. and Duffey, K.A. (1985) *The failure to demonstrate avoidance of ventral gland odors in male gerbils* (Meriones unguiculatus). Physiol. Behav., 35, 763–765.

Furudate, S. and Nakano, T. (1981) *[Studies on the pheromonal pregnancy block in the mouse. II. Discriminatory and/or memorial mechanisms in female mice responding to the stud male].* [Japanese]. Jikken Dobutsu, 30, 1–5.

Galef, B.G., Jr. (1981) *Preference for natural odors in rat pups: implications of a failure to replicate.* Physiol. Behav., 26, 783–786.

Galef, B.G., Jr. (1986) *Social identification of toxic diets by Norway rats* (Rattus norvegicus). J. Comp. Psychol., 100, 331–334.

Galef, B.G., Jr., Beck, M. and Whiskin, E.E. (1991) *Protein deficiency magnifies social influence on the food choices of Norway rats* (Rattus norvegicus). J. Comp. Psychol., 105, 55–59.

Galef, B.G., Jr. and Henderson, P.W. (1972) *Mother's milk: a determinant of the feeding preferences of weaning rat pups.* J. Comp. Physiol. Psychol., 78, 213–219.

Galef, B.G., Jr., Iliffe, C.P. and Whiskin, E.E. (1994) *Social influences on rats'* (Rattus norvegicus) *preferences for flavored foods, scented nest materials,* and odors associated with harborage sites: are flavored foods special? J. Comp. Psychol., 108, 266–273.

Galef, B.G., Jr. and Kaner, H.C. (1980) *Establishment and maintenance of preference for natural and artificial olfactory stimuli in juvenile rats.* J. Comp. Physiol. Psychol., 94, 588–595.

Galef, B.G., Jr. and Sherry, D.F. (1973) *Mother's milk: a medium for transmission of cues reflecting the flavor of mother's diet.* J. Comp. Physiol. Psychol., 83, 374–378.

Galef, B.G., Jr., Whiskin, E.E. and Bielavska, E. (1997) *Interaction with demonstrator rats changes observer rats' affective responses to flavors.* J. Comp. Psychol., 111, 393–398.

Garcia-Velasco, J. and Mondragon, M. (1991) *The incidence of the vomeronasal organ in 1000 human subjects and its possible clinical significance.* J. Steroid Biochem. Mol. Biol., 39, 561–563.

Gelez, H., Archer, E., Chesneau, D., Campan, R. and Fabre-Nys, C. (2004a) *Importance of learning in the response of ewes to male odor.* Chem. Senses, 29, 555–563.

Gelez, H., Archer, E., Chesneau, D., Magallon, T. and Fabre-Nys, C. (2004b) *Inactivation of the olfactory amygdala prevents the endocrine response to male odour in anoestrus ewes.* Eur. J. Neurosci., 19, 1581–1590.

Giersing, M., Lundstrom, K. and Andersson, A. (2000) *Social effects and boar taint: significance for production of slaughter boars* (Sus scrofa). J. Anim. Sci., 78, 296–305.

Gleason, K.K. and Reynierse, J.H. (1969) *The behavioral significance of pheromones in vertebrates.* Psychol. Bull., 71, 58–73.

Goldfoot, D.A., Essock-Vitale, S.M., Asa, C.S., Thornton, J.E. and Leshner, A.I. (1978) *Anosmia in male rhesus monkeys does not alter copulatory activity with cycling females.* Science, 199, 1095–1096.

Goldfoot, D.A., Kravetz, M.A., Goy, R.W. and Freeman, S.K. (1976) *Lack of effect of vaginal lavages and aliphatic acids on ejaculatory responses in rhesus monkeys: behavioral and chemical analyses.* Horm. Behav., 7, 1–27.

Goldman, S.E. and Schneider, H.G. (1987) *Menstrual synchrony: social and personality factors.* J. Soc. Behav. Personal., 2, 243–250.

Gonzalez-Bono, E., Salvador, A., Serrano, M.A. and Ricarte, J. (1999) *Testosterone, cortisol, and mood in a sports team competition.* Horm. Behav., 35, 55–62.

Gonzalez-Mariscal, G., Chirino, R. and Hudson, R. (1994) *Prolactin stimulates emission of nipple pheromone in ovariectomized New Zealand white rabbits.* Biol. Reprod., 50, 373–376.

Goodwin, M., Gooding, K.M. and Regnier, F. (1979) *Sex pheromone in the dog.* Science, 203, 559–561.

Gosling, L.M. (1982) *A reassessment of the function of scent marking in territories*. Z. Tierpsychol., 60, 89–118.

Gosling, L.M. (1990) *Scent marking by resource holders: alternative mechanisms for advertising the costs of competition*. In Macdonald, D., Muller-Schwarze, D. and Natynczuk, S.E. (eds.), Chemical signals in vertebrates 5. Oxford University Press, Oxford, pp. 315–328.

Gosling, L.M., Atkinson, N.W., Collins, S.A., Roberts, R.J. and Walters, R.L. (1996a) *Avoidance of scent-marked areas depends on the intruder's body size*. Behaviour, 133, 491–502.

Gosling, L.M., Atkinson, N.W., Dunn, S. and Collins, S.A. (1996b) *The response of subordinate male mice to scent marks varies in relation to their own competitive ability*. Anim. Behav., 52, 1185–1191.

Gosling, L.M. and McKay, H.V. (1990) *Competitor assessment by scent matching: an experimental test*. Behav. Ecol. Sociobiol., 26, 415–420.

Gosling, L.M. and Roberts, S.C. (2001) *Scent marking by male mammals: cheat-proof signals to competitors and mates*. Adv. Study Beh., 30, 169–217.

Gower, D.B., Bird, S., Sharma, P. and House, F.R. (1985) *Axillary 5 α-androst-16-en-3-one in men and women: relationships with olfactory acuity to odorous 16-androstenes*. Experientia, 41, 1134–1136.

Gower, D.B., Holland, K.T., Mallet, A.I., Rennie, P.J. and Watkins, W.J. (1994) *Comparison of 16-androstene steroid concentrations in sterile apocrine sweat and axillary secretions: interconversions of 16-androstenes by the axillary microflora—a mechanism for axillary odour production in man?* J. Steroid Biochem. Mol. Biol., 48, 409–418.

Gower, D.B., Nixon, A. and Mallet, A.I. (1988) *The significance of odorous steroids in axillary odor*. In Van Toller, S. and Dodd, G.H. (eds.), Perfumery: the psychology and biology of fragrance. Chapman & Hall, London, pp. 47–76.

Gower, D.B. and Ruparelia, B.A. (1993) *Olfaction in humans with special reference to odorous 16-androstenes: their occurrence, perception and possible social, psychological and sexual impact*. J. Endocrinol., 137, 167–187.

Goyens, J. and Noirot, E. (1975) *Effects of cohabitation with females on aggressive behavior between male mice*. Dev Psychobiol., 8, 79–84.

Graham, C.A., Janssen, E. and Sanders, S.A. (2000) *Effects of fragrance on female sexual arousal and mood across the menstrual cycle*. Psychophysiology, 37, 76–84.

Graham, C.A. and McGrew, W.C. (1980) *Menstrual synchrony in female undergraduates living on a coeducational campus*. Psychoneuroendocrinology, 5, 245–252.

Graham, J.M. and Desjardins, C. (1980) *Classical conditioning: induction of luteinizing hormone and testosterone secretion in anticipation of sexual activity*. Science, 210, 1039–1041.

Grammer, K., Fink, B. and Neave, N. (2005) *Human pheromones and sexual attraction.* Eur. J. Obst. Gynecol. Repr. Biol., 118, 135–142.

Greenwood, D.R., Comeskey, D., Hunt, M.B. and Rasmussen, L.E.L. (2005) *Chemical communication—chirality in elephant pheromones.* Nature, 438, 1097–1098.

Gregoire, A.T., Ledger, W.D. and Moran, M.J. (1973) *The glycogen content of the human female genital tract in cycling, menopausal and women with endometrial and cervical carcinoma.* Fertil. Steril., 24, 198–201.

Gregory, E.H. and Bishop, A. (1975) *Development of olfactory-guided behavior in the golden hamster.* Physiol. Behav., 15, 373–376.

Grigoriadis, D.E., Struble, R.G., Price, D.L. and De Souza, E.B. (1989) *Normal pattern of labeling of cerebral cortical corticotropin-releasing factor (CRF) receptors in Alzheimer's disease: evidence from chemical cross-linking studies.* Neuropharmacology, 28, 761–764.

Grosser, B.I., Monti-Bloch, L., Jennings-White, C. and Berliner, D.L. (2000) *Behavioral and electrophysiological effects of androstadienone, a human pheromone.* Psychoneuroendocrinology, 25, 289–299.

Groves, P. and Schlesinger, K. (1979) Introduction to biological psychology. William C. Brown Company, Dubuque, IA.

Grüneberg, H. (1973) *A ganglion probably belonging to the N. terminalis system in the nasal mucosa of the mouse.* Z. Anat. Entwicklungsgesch., 140, 39–52.

Gustavson, A.R., Dawson, M.E. and Bonett, D.G. (1987) *Androstenol, a putative human pheromone, affects human* (Homo sapiens) *male choice performance.* J. Comp. Psychol., 101, 210–212.

Haertzen, C. (1974) Addiction Research Inventory. National Institutes of Mental Health, Washington, DC.

Hafez, E.S.E. and Signoret, J.P. (1969) *The behaviour of swine.* In Hafez, E.S.E. (ed.), The behavior of domestic animals. Williams and Wilkins, Baltimore, pp. 349–390.

Handelmann, G., Ravizza, R. and Ray, W.J. (1980) *Social dominance determines estrous entrainment among female hamsters.* Horm. Behav., 14, 107–115.

Haring, H.G. (1974) *Vapor pressures and Raoult's law deviations in relation to odor enhancement and suppression.* In Turk, A., Johnston, J.W., Jr. and Moulton, D.G. (eds.), Human responses to environmental odors. Academic Press, New York, pp. 199–226.

Harlow, S.D. and Zeger, S.L. (1991) *An application of longitudinal methods to the analysis of menstrual diary data.* J. Clin. Epidemiol., 44, 1015–1025.

Harrington, F.H. (1981) *Urine-marking and caching behaviour in the wolf.* Behaviour, 76, 280–288.

Hartmann, M. and Schartau, O. (1939) *Untersuchungen über die Befruchtungsstoffe der Seeigel I.* Biol. Zentralblatt., 59, 571–587.

Hasegawa, Y., Yabuki, M. and Matsukane, M. (2004) *Identification of new odoriferous compounds in human axillary sweat.* Chem. Biodivers., 1, 2042–2050.

Hatanaka, T. (1992) *Is the mouse vomeronasal organ a sex pheromone receptor?* In Doty, R.L. and Müller-Schwarze, D. (eds.), Chemical signals in vertebrates 6. Plenum Press, New York, pp. 27–30.

Hatanaka, T., Shibuya, T. and Inouchi, J. (1988) *Induced wave responses of the accessory olfactory bulb to odorants in two species of turtle,* Pseudemys scripta *and* Geoclemys reevesii. Comp. Biochem. Physiol., 91A, 377–385.

Hatch, M.C., Figa-Talamanca, I. and Salerno, S. (1999) *Work stress and menstrual patterns among American and Italian nurses.* Scand. J. Work. Environ. Health, 25, 144–150.

Hauser, R., Marczak, M., Namiesnik, J. and Karaszewski, B. (2005) *A chamber for testing the release of volatile substances secreted by animals, especially mammals, as exemplified by substances released by rats in response to stress.* Instr. Sci. Tech., 33, 541–550.

Havlicek, J. and Lenochova, P. (2006) *The effect of meat consumption on body odor attractiveness.* Chem. Senses, 31, 747–752.

Hayashi, S. (1979) *A role of female preputial glands in social behavior of mice.* Physiol. Behav., 23, 967–969.

Hayashi, S. and Kimura, T. (1974) *Sex-attractant emitted by female mice.* Physiol. Behav., 13, 563–567.

Hayashi, S. and Kimura, T. (1976) *Sexual behavior of the naive male mouse as affected by the presence of a male and a female performing mating behavior.* Physiol. Behav., 17, 807–810.

Heise, S. and Hurst, J.L. (1994) *Territorial experience causes a shift in the responsiveness of female house mice to odours from dominant males.* Adv. Biosci., 93, 291–296.

Hellhammer, D.H., Hubert, W. and Schurmeyer, T. (1985) *Changes in saliva testosterone after psychological stimulation in men.* Psychoneuroendocrinology, 10, 77–81.

Helmreich, R.L. (1960) *Regulation of reproductive rate by intra-uterine mortality in the deer mouse.* Science, 132, 417–418.

Hendry, L.B., Wichmann, J.K., Hindenlang, D.M., Mumma, R.O. and Anderson, M.E. (1975) *Evidence for origin of insect sex pheromones: presence in food plants.* Science, 188, 59–63.

Henry, J.D. (1977) *The use of urine marking in the scavenging behavior of the red fox* (Vulpes vulpes). Behaviour, 61, 82–106.

Hepper, P.G. (1987) *The amniotic fluid: an important priming role in kin recognition.* Anim. Behav., 35, 1343–1346.

Hepper, P.G. (1988) *Adaptive fetal learning: prenatal exposure to garlic affects postnatal preferences.* Anim. Behav., 36, 935–936.

Heyser, C.J., Spear, N.E. and Spear, L.P. (1992) *Effects of prenatal exposure to cocaine on conditional discrimination learning in adult rats.* Behav. Neurosci., 106, 837–845.

Hill, E.M. (1988) *The menstrual cycle and components of human female sexual behavior.* J. Soc. Biol. Struct., 11, 443–455.

Hodos, W. and Campbell, C.B.G. (1969) *Scala naturae: why there is no theory in comparative psychology.* Psychol. Bull., 76, 337–350.

Hoffmann, G., Gufler, V., Griesmacher, A., Bartenbach, C., Canazei, M., Staggl, S. and Schobersberger, W. (2008) *Effects of variable lighting intensities and colour temperatures on sulphatoxymelatonin and subjective mood in an experimental office workplace.* Appl. Ergonom., 39, 719–728.

Hold, B. and Schleidt, M. (1977) *The importance of human odour in non-verbal communication.* Z. Tierpsychol., 43, 225–238.

Holinka, C.F. and Carlson, A.D. (1976) *Pup attraction to lactating Sprague-Dawley rats.* Behav. Biol., 16, 489–505.

Holldobler, B. (1999) *Multimodal signals in ant communication.* J. Comp. Physiol. A, 184, 129–141.

Holldobler, B. and Carlin, N.F. (1987) *Anonymity and specificity in the chemical communication signals of social insects.* J. Comp. Physiol. A, 161, 567–581.

Holy, T.E., Dulac, C. and Meister, M. (2000) *Responses of vomeronasal neurons to natural stimuli.* Science, 289, 1569–1572.

Hopp, S.L. and Timberlake, W. (1983) *Odor cue determinants of urine marking in male rats* (Rattus norvegicus). Behav. Neur. Biol., 37, 162–172.

Hoppe, P.C. (1975) *Genetic and endocrine studies of the pregnancy-blocking pheromone of mice.* J. Reprod. Fertil., 45, 109–115.

Hoppe, R., Breer, H. and Strotmann, J. (2003) *Organization and evolutionary relatedness of OR37 olfactory receptor genes in mouse and human.* Genomics, 82, 355–364.

Huang, H.H., Marshall, S. and Meites, J. (1976) *Induction of estrous cycles in old non-cyclic rats by progesterone, ACTH, ether stress or L-Dopa.* Neuroendocrinology, 20, 21–34.

Huck, U.W. and Banks, E.M. (1984) *Social olfaction in male brown lemmings* (Lemmus sibiricus = trimucronatus) *and collared lemmings* (Dicrostonyx groenlandicus): *I. Discrimination of species, sex, and estrous condition.* J. Comp. Psychol., 98, 54–59.

Hudson, R. (1999) *From molecule to mind: the role of experience in shaping olfactory function.* J. Comp. Physiol. A, 185, 297–304.

Hudson, R. and Altbäcker, V. (1982) *Development of feeding and food preference in the European rabbit: environmental and maturational determinants.* In Galef, B.G., Mainardi, M. and Valsecchu, P. (eds.), Behavioral aspects of feeding: basic and applied research in mammals. Harood Academic Publishers, Newark, pp. 125–145.

Hudson, R. and Distel, H. (1983) *Nipple location by newborn rabbits: behavioural evidence for pheromonal guidance.* Behaviour, 85, 260–275.

Hudson, R. and Distel, H. (1986) *Pheromonal release of suckling in rabbits does not depend on the vomeronasal organ.* Physiol. Behav., 37, 123–128.

Hudson, R. and Distel, H. (1999) *Induced peripheral sensitivity in the developing vertebrate olfactory system.* Ann. NY Acad. Sci., 856, 109–115.

Hudson, R., Labra-Cardero, D. and Mendoza-Soylovna, A. (2002) *Sucking, not milk, is important for the rapid learning of nipple-search odors in newborn rabbits.* Dev. Psychobiol., 41, 226–235.

Hudson, R., Rojas, C., Arteaga, L., Martinez-Gomez, M. and Distel, H. (2007) *Rabbit nipple-search pheromone versus rabbit mammary pheromone revisited.* In Hurst, J. and Benyon, R. (eds.), Chemical signals in vertebrates 11. Springer, New York, pp. 315–324.

Huggins, G.R. and Preti, G. (1976) *Volatile constituents of human vaginal secretions.* Am. J. Obstet. Gynecol., 126, 129–136.

Huh, J., Park, K., Hwang, I.S., Jung, S.I., Kim, H.J., Chung, T.W. and Jeong, G.W. (2008) *Brain activation areas of sexual arousal with olfactory stimulation in men: a preliminary study using functional MRI.* J. Sex. Med., 5, 619–625.

Hummel, T. and Welge-Lussen, A.E. (2006) Taste and smell: an update. Karger, Basel.

Hurst, J.L. (1993) *The priming effects of urine substrate marks on interactions between male house mice,* Mus musculus domesticus, Schwartz and Schwartz. Anim. Behav., 45, 55–81.

Hurst, J.L., Hayden, L., Kingston, M., Luck, R. and Sorensen, K. (1994) *Response of the aboriginal house mouse* Mus spretus Lataste *to tunnels bearing the odours of conspecifics.* Anim. Behav., 48, 1219–1229.

Hurst, J.L., Payne, C.E., Nevison, C.M., Marie, A.D., Humphries, R.E., Robertson, D.H., Cavaggioni, A. and Beynon, R.J. (2001) *Individual recognition in mice mediated by major urinary proteins.* Nature, 414, 631–634.

Hurst, J.L., Thom, M.D., Nevison, C.M., Humphries, R.E. and Beynon, R.J. (2005) *MHC odours are not required or sufficient for recognition of individual scent owners.* Proc. Roy. Soc. Lond. B: Biol. Sci., 272, 715–724.

Husain, G., Thompson, W.F. and Schellenberg, E.G. (2002) *Effects of musical tempo and mode on arousal, mood, and spatial abilities.* Music Percept., 20, 151–171.

Idris, M. and Prakash, I. (1986) *Influence of odours of Indian gerbil,* Tatera indica, *on the social and scent marking behaviour of sympatric desert gerbil,* Meriones hurrianae. Proc. Indian Nat. Sci. Acad., B52, 333–340.

Izard, C.E., Libero, D.Z., Putnam, P. and Haynes, O.M. (1993) *Stability of emotion experiences and their relations to traits of personality.* J. Pers. Soc. Psychol., 64, 847–860.

Izard, M.K. and Vandenbergh, J.G. (1982) *Priming pheromones from oestrous cows increase synchronization of oestrus in dairy heifers after PGF-2 α injection.* J. Reprod. Fertil., 66, 189–196.

Jacob, S., Garcia, S., Hayreh, D. and McClintock, M.K. (2002) *Psychological effects of musky compounds: comparison of androstadienone with androstenol and muscone.* Horm. Behav., 42, 274–283.

Jacob, S., Hayreh, D.J. and McClintock, M.K. (2001) *Context-dependent effects of steroid chemosignals on human physiology and mood.* Physiol. Behav., 74, 15–27.

Jacob, S. and McClintock, M.K. (2000) *Psychological state and mood effects of steroidal chemosignals in women and men.* Horm. Behav., 37, 57–78.

Jacob, S., Spencer, N.A., Bullivant, S.B., Sellergren, S.A., Mennella, J.A. and McClintock, M.K. (2004) *Effects of breastfeeding chemosignals on the human menstrual cycle.* Hum. Reprod., 19, 422–429.

Jacob, S., Zelano, B., Gungor, A., Abbott, D., Naclerio, R. and McClintock, M.K. (2000) *Location and gross morphology of the nasopalatine duct in human adults.* Arch. Otolaryngol. Head Neck Surg., 126, 741–748.

Jacob, T.J.C., Wang, L.W., Jaffer, S. and McPhee, S. (2006) *Changes in the odor quality of androstadienone during exposure-induced sensitization.* Chem. Senses, 31, 3–8.

Jahn, G.A. and Deis, R.P. (1986) *Stress-induced prolactin release in female, male and androgenized rats: influence of progesterone treatment.* J. Endocrinol., 110, 423–428.

Janeczko, A. and Skoczowski, A. (2005) *Mammalian sex hormones in plants.* Fol. Histochem. Cytobiol., 43, 71–79.

Jang, T., Singer, A.G. and O'Connell, R.J. (2001) *Induction of c-fos in hamster accessory olfactory bulbs by natural and cloned aphrodisin.* Neuroreport, 12, 449–452.

Janus, C. (1989) *The development of olfactory preferences for artificial odors briefly experienced by the precocial spiny mouse young.* Behav. Neur. Biol., 52, 430–436.

Janus, C. (1993) *Stability of preference for odors after short-term exposure in young spiny mice.* Dev. Psychobiol., 26, 65–79.

Jarett, L.R. (1984) *Psychosocial and biological influences on menstruation: synchrony, cycle length, and regularity.* Psychoneuroendocrinology, 9, 21–28.

Jemiolo, B., Andreolini, F., Xie, T.M., Wiesler, D. and Novotny, M. (1989) *Puberty-affecting synthetic analogs of urinary chemosignals in the house mouse*, Mus domesticus. Physiol. Behav., 46, 293–298.

Jennings, J.W. and Keffer, L.H. (1969) *Olfactory learning set in two varieties of domestic rat*. Psychol. Rep., 24, 3–15.

Johnson, B.A. and Leon, M. (2007) *Chemotropic odorant coding in a mammalian olfactory system*. J. Comp. Neurol., 503, 1–34.

Johnston, R.E. (1975) *Sexual excitation function of hamster vaginal secretion*. Anim. Learn. Behav., 3, 161–166.

Johnston, R.E. (1993) *Memory for individual scent in hamsters (*Mesocricetus auratus) *as assessed by habituation methods*. J. Comp. Psychol., 107, 201–207.

Johnston, R.E. (1999) *Scent over-marking. How do hamsters know whose scent is on top and why should it matter?* In Johnston, R.E., Müller-Schwarze, D. and Sorensen, P.W. (eds.), Advances in chemical signals in vertebrates. Kluwer Academic/Plenum Publishers, New York, pp. 227–238.

Johnston, R.E. (2000) *Chemical communication and pheromones: the types of chemical signals and the role of the vomeronasal system*. In Finger, T.E., Silver, W.L. and Restrepo, D. (eds.), The neurobiology of taste and smell. Wiley-Liss, New York, pp. 101–127.

Johnston, R.E. and Bhorade, A. (1998) *Perception of scent over-marks by golden hamsters* (Mesocricetus auratus): *novel mechanisms for determining which individual's mark is on top*. J. Comp. Psychol., 112, 230–243.

Johnston, R.E. and Bullock, T.A. (2001) *Individual recognition by use of odours in golden hamsters: the nature of individual representations*. Anim. Behav., 61, 545–557.

Johnston, R.E. and Coplin, B. (1979) *Development of responses to vaginal secretion and other substances in golden hamsters*. Behav. Neur. Biol., 25, 473–489.

Johnston, R.E. and Jernigan, P. (1994) *Golden hamsters recognize individuals, not just individual scents*. Anim. Behav., 48, 129–136.

Johnston, R.E. and Peng, A. (2008) *Memory for individuals: hamsters* (Mesocricetus auratus) *require contact to develop multicomponent representations (concepts) of others*. J. Comp. Psychol., 122, 121–131.

Johnston, R.E. and Zahorik, D.M. (1975) *Taste aversions to sexual attractants*. Science, 189, 893–894.

Jones, R.B. and Nowell, N.W. (1973) *The effect of urine on the investigatory behaviour of male albino mice*. Physiol. Behav., 11, 35–38.

Jones, R.B. and Nowell, N.W. (1974a) *Effects of cyproterone acetate upon urinary aversive cues and accessory sex glands in male albino mice*. J. Endocrinol., 62, 167–168.

Jones, R.B. and Nowell, N.W. (1974b) *A comparison of the aversive and female attractant properties of urine from dominant and subordinate male mice.* Med. Weterynaryjna, 2, 141–144.

Jones, R.B. and Nowell, N.W. (1989) *Aversive potency of urine from dominant and subordinate male laboratory mice* (Mus musculus): *resolution of a conflict.* Aggressive Beh., 15, 291–296.

Jurtshuk, P., Weltman, A.S. and Sackler, A.M. (1959) *Biochemical responses of rats to auditory stress.* Science, 129, 1424–1425.

Kaba, H. and Keverne, E.B. (1988) *The effect of microinfusions of drugs into the accessory olfactory bulb on the olfactory block to pregnancy.* Neurosci., 25, 1007–1011.

Kaba, H., Rosser, A. and Keverne, B. (1989) *Neural basis of olfactory memory in the context of pregnancy block.* Neurosci., 32, 657–662.

Kakihana, R., Ellis, L.B., Gerling, S.A., Blum, S.L. and Kessler, S. (1974) *Bruce effect competence in yellow-lethal heterozygous mice.* J. Repr. Fertil., 40, 483–486.

Karev, G.B. (2000) *Cinema seating in right, mixed and left handers.* Cortex, 36, 747–752.

Karlson, P. and Butenandt, A. (1959) *Pheromones (ectohormones) in insects.* Ann. Rev. Entomol., 4, 39–58.

Karlson, P. and Lüscher, M. (1959) *"Pheromones": a new term for a class of biologically active substances.* Nature, 183, 55–56.

Katz, D.B., Matsunami, H., Rinberg, D., Scott, K., Wachowiak, M. and Wilson, R.I. (2008) *Receptors, circuits, and behaviors: new directions in chemical senses.* J. Neurosci., 28, 11802–11805.

Katz, R.A. and Shorey, H.H. (1979) *In defense of the term "pheromone."* J. Chem. Ecol., 5, 299–301.

Kelley, R.B. (1937) Studies in fertility of sheep. H.J. Green, Melbourne.

Kelliher, K.R., Spehr, M., Li, X.H., Zufall, F. and Leinders-Zufall, T. (2006) *Pheromonal recognition memory induced by TRPC2-independent vomeronasal sensing.* Eur. J. Neurosci., 23, 3385–3390.

Kelly, J.P., Wrynn, A.S., Leonard, B.E., Kelly, J.P., Wrynn, A.S. and Leonard, B.E. (1997) *The olfactory bulbectomized rat as a model of depression: an update.* Pharmacol. Therap., 74, 299–316.

Kendrick, K. (1975) *Maternal pheromone: discrimination by pre-weanling rats.* Unpublished dissertation, University of Durham, Durham, England.

Kendrick, K.M., Haupt, M.A., Hinton, M.R., Broad, K.D. and Skinner, J.E. (2001) *Sex differences in the influence of mothers on the sociosexual preferences of their offspring.* Horm. Behav., 40, 322–338.

Kendrick, K.M., Levy, F., and Keverne, E.B. (1992) *Changes in the sensory processing of olfactory signals induced by birth in sheep.* Science, 256, 833–836.

Kerchner, M., Vatza, E.J. and Nyby, J. (1986) *Ultrasonic vocalizations by male house mice* (Mus musculus) *to novel odors: roles of infant and adult experience.* J. Comp. Psychol., 100, 253–261.

Keverne, E.B. (1974) *Sex-attractants in primates.* New Scientist, 51, 313–322.

Keverne, E.B. and de la Riva, C. (1982) *Pheromones in mice: reciprocal interaction between the nose and brain.* Nature, 296, 148–150.

Keverne, E.B. and Michael, R.P. (1971) *Sex-attractant properties of ether extracts of vaginal secretions from rhesus monkeys.* J. Endocrinol., 51, 313–322.

Kiecolt-Glaser, J.K., Graham, J.E., Malarkey, W.B., Porter, K., Lemeshow, S. and Glaser, R. (2008) *Olfactory influences on mood and autonomic, endocrine, and immune function.* Psychoneuroendocrinology, 33, 328–339.

Kikusui, T., Takigami, S., Takeuchi, Y. and Mori, Y. (2001) *Alarm pheromone enhances stress-induced hyperthermia in rats.* Physiol. Behav., 72, 45–50.

Kimelman, B.R. and Lubow, R.E. (1974) *The inhibitory effect of preexposed olfactory cues on intermale aggression in mice.* Physiol. Behav., 12, 919–922.

Kindermann, U., Gervais, R. and Hudson, R. (1991) *Rapid odor conditioning in newborn rabbits: amnesic effect of hypothermia.* Physiol. Behav., 50, 457–460.

Kindermann, U., Hudson, R. and Distel, H. (1994) *Learning of suckling odors by newborn rabbits declines with age and suckling experience.* Dev Psychobiol., 27, 111–122.

King, J.A. (1959) *Effects of early handling upon adult behavior in two subspecies of deermice,* Permyscus maniculatus. J. Comp. Physiol. Psychol., 52, 82–88.

King, J.A. (1973) *The ecology of aggressive behavior.* Ann. Rev. Ecol. Evol. Syst., 4, 117–138.

King, M.G., Pfister, H.P. and DiGiusto, E.L. (1975) *Differential preference for and activation by the odoriferous compartment of a shuttlebox in fear-conditioned and naive rats.* Behav. Biol., 13, 175–181.

Kippin, T.E., Cain, S.W. and Pfaus, J.G. (2003) *Estrous odors and sexually conditioned neutral odors activate separate neural pathways in the male rat.* Neurosci., 117, 971–979.

Kippin, T.E. and Pfaus, J.G. (2001a) *Nature of the conditioned response mediating olfactory conditioned ejaculatory preference in the male rat.* Behav. Brain Res., 122, 11–24.

Kippin, T.E. and Pfaus, J.G. (2001b) *The development of olfactory conditioned ejaculatory preferences in the male rat. I. Nature of the unconditioned stimulus.* Physiol. Behav., 73, 457–469.

Kirk-Smith, M.D. and Booth, D.A. (1980) *Effect of androstenone on choice of location in other's presence.* In van der Starre, H. (ed.), Olfaction and taste VII. IRL Press, London, pp. 397–400.

Kirk-Smith, M., Booth, D.A., Carroll, D. and Davies, P. (1978) *Human social attitudes affected by androstenol.* Res. Comm. Psychol. Psychiat. Behav., 3, 379–384.

Kiyokawa, Y., Kikusui, T., Takeuchi, Y. and Mori, Y. (2004a) *Alarm pheromones with different functions are released from different regions of the body surface of male rats.* Chem. Senses, 29, 35–40.

Kiyokawa, Y., Kikusui, T., Takeuchi, Y. and Mori, Y. (2004b) *Modulatory role of testosterone in alarm pheromone release by male rats.* Horm. Behav., 45, 122–127.

Kiyokawa, Y., Kikusui, T., Takeuchi, Y. and Mori, Y. (2005) *Alarm pheromone that aggravates stress-induced hyperthermia is soluble in water.* Chem. Senses, 30, 513–519.

Kloek, J. (1961) *The smell of some steroid sex-hormones and their metabolites: reflections and experiments concerning the significance of smell for the mutual relation of the sexes.* Psychiat. Neurol. Neurochir., 64, 309–344.

Knecht, M., Kuhnau, D., Huttenbrink, K.B., Witt, M. and Hummel, T. (2001) *Frequency and localization of the putative vomeronasal organ in humans in relation to age and gender.* Laryngoscope, 111, 448–452.

Knight, T.W. and Lynch, P.R. (1980) *Source of ram pheromones that stimulate ovulation in the ewe.* Anim. Reprod. Sci., 3, 133–136.

Knutson, B., Burgdorf, J. and Panksepp, J. (1998) *Anticipation of play elicits high-frequency ultrasonic vocalizations in young rats.* J. Comp. Psychol., 112, 65–73.

Knutson, B., Burgdorf, J. and Panksepp, J. (1999) *High-frequency ultrasonic vocalizations index conditioned pharmacological reward in rats.* Physiol. Behav., 66, 639–643.

Kodis, M., Moran, D. and Houy, D. (1998) Love scents: how your natural pheromones influence your relationships, your moods, and who you love. Dutton, New York.

Koelega, H.S. (1980) *Preference for and sensitivity to the odours of androstenone and musk.* In van der Starre, H. (ed.), Proceedings of the 7th International Symposium on Olfaction & Taste and the 4th Congress of the ECRO. IRL Press, London, p. 436.

Kollack-Walker, S. and Newman, S.W. (1997) *Mating-induced expression of c-fos in the male Syrian hamster brain: role of experience, pheromones, and ejaculations.* J. Neurobiol., 32, 481–501.

Kondo, Y., Sachs, B.D. and Sakuma, Y. (1998) *Importance of the medial amygdala in rat penile erection evoked by remote stimuli from estrous females.* Behav. Brain Res., 91, 215–222.

Kondo, Y., Tomihara, K. and Sakuma, Y. (1999) *Sensory requirements for noncontact penile erection in the rat.* Behav. Neurosci., 113, 1062–1070.

Kongsted, A.G. (2004) *Stress and fear as possible mediators of reproduction problems in group housed sows: a review.* Acta Agri. Scand. A. Anim. Sci., 54, 58–66.

Kouros-Mehr, H., Pintchovski, S., Melnyk, J., Chen, Y.J., Friedman, C., Trask, B. and Shizuya, H. (2001) *Identification of non-functional human VNO receptor genes provides evidence for vestigiality of the human VNO.* Chem. Senses, 26, 1167–1174.

Kroner, C., Breer, H., Singer, A.G. and O'Connell, R.J. (1996) *Pheromone-induced second messenger signaling in the hamster vomeronasal organ.* Neuroreport, 7, 2989–2992.

Krout, K.E., Kawano, J., Mettenleiter, T.C. and Loewy, A.D. (2002) *CNS inputs to the suprachiasmatic nucleus of the rat.* Neurosci., 110, 73–92.

Kruse, S.M. and Howard, W.E. (1983) *Canid sex attractant studies.* J. Chem. Ecol., 9, 1503–1510.

Krutova, V.I. and Zinkevich, E.P. (1999) *Gray rats* (Rattus norvegicus) *memorize without training and distinguish human scents.* Dokl. Biol. Sci. Akad. Nauk SSSR, 369, 573–575.

Krzymowski, T., Grzegorzewski, W., Stefanczyk-Krzymowska, S., Skipor, J. and Wasowska, B. (1999) *Humoral pathway for transfer of the boar pheromone, androstenol, from the nasal mucosa to the brain and hypophysis of gilts.* Theriogenology, 52, 1225–1240.

Kurt, F. (1967) *Zur Rolle des Geruchs im Verhalten des Rehwildes.* Verh. Schweiz. Naturforsch. Ges. Zürich., 144, 140–142.

Kwan, T.K., Kraevskaya, M.A., Makin, H.L., Trafford, D.J. and Gower, D.B. (1997) *Use of gas chromatographic-mass spectrometric techniques in studies of androst-16-ene and androgen biosynthesis in human testis; cytosolic specific binding of 5α-androst-16-en-3-one.* J. Steroid Biochem. Mol. Biol., 60, 137–146.

Kwan, T.K., Trafford, D.J., Makin, H.L., Mallet, A.I. and Gower, D.B. (1992) *GC-MS studies of 16-androstenes and other C19 steroids in human semen.* J. Steroid Biochem. Mol. Biol., 43, 549–556.

Labov, J.B. (1981) *Pregnancy blocking in rodents: adaptive advantages for females.* Amer. Naturalist, 118, 361–371.

Labows, J. (1988) *Odor detection, generation and etiology in the axilla.* In Felger, C. and Laden, K. (eds.), Antiperspirants and deodorants. Marcel Dekker, New York, pp. 321–343.

Labows, J.N., Preti, G., Hoelzle, E., Leyden, J. and Kligman, A. (1979) *Steroid analysis of human apocrine secretion.* Steroids, 34, 249–258.

Lalumiere, M.L., Blanchard, R. and Zucker, K.J. (2000) *Sexual orientation and handedness in men and women: a meta-analysis.* Psychol. Bull., 126, 575–592.

Lamond, D.R. (1958a) *Infertility associated with extirpation of the olfactory bulbs in female albino mice.* Aust. J. Exp. Biol. Med. Sci., 36, 103–108.

Lamond, D.R. (1958b) *Spontaneous anoestrus in mice.* Proc. Australian Soc. Animal. Prod., 2, 97–101.

Lamond, D.R. (1959) *Effect of stimulation derived from other animals of the same species on oestrus cycles in mice.* J. Endocrinol., 18, 343–349.

Lang, P.J., Bradley, M.M. and Cuthbert, B.N. (1999) International Affective Picture System (IAPS): Technical Manual and Affective Ratings. The Center for Research in Psychophysiology, University of Florida, Gainesville, Florida.

Larson, J. and Sieprawska, D. (2002) *Automated study of simultaneous-cue olfactory discrimination learning in adult mice.* Behav. Neurosci., 116, 588–599.

Lawless, H. (1991) *Effects of odors on mood and behavior: aromatherapy and related effects.* In Laing, D.G., Doty, R.L. and Breipohl, W. (eds.), The human sense of smell. Springer-Verlag, Berlin, pp. 361–386.

Le Magnen, J. (1952a) *Les phénomènes olfacto-sexuels chez le rat blanc.* Arch. Sci. Physiol., 6, 295–332.

Le Magnen, J. (1952b) *Les phenomenes olfacto-sexuels chez l'homme.* C. R. Acad. Sci. Biol., 6, 125–160.

Lee, C.T. and Brake, S.C. (1971) *Reactions of male fighters to male and female mice, untreated or deodorized.* Psychonomic. Sci., 24, 209–211.

Lee, C.T. and Griffo, W. (1973) *Early androgenization and aggression pheromone in inbred mice.* Horm. Behav., 4, 181–189.

Lee, C.T. and Griffo, W. (1974) *Progesterone antagonism of androgen-dependent aggression-promoting pheromone in inbred mice* (Mus musculus). J. Comp. Physiol. Psychol., 87, 150–155.

Lehrman, D.S. (1953) *A critique of Konrad Lorenz' theory of instinctive behavior.* Quart. Rev. Biol., 28, 337–363.

Leidahl, L.C. and Moltz, H. (1975) *Emission of the maternal pheromone in the nulliparous female and failure of emission in the adult male.* Physiol. Behav., 14, 421–424.

Lent, P. (1966) *Calving and related social behavior in the barren-ground caribou.* Z. Tierpsychol., 23, 702–756.

Leon, M. (1974) *Maternal pheromone.* Physiol. Behav., 13, 441–453.

Leon, M. (1975) *Dietary control of maternal pheromone in the lactating rat.* Physiol. Behav., 14, 311–319.

Leon, M. (1978) *Emission of maternal pheromone.* Science, 201, 938–939.

Leon, M. and Moltz, H. (1971) *Maternal pheromone: discrimination by pre-weanling albino rats.* Physiol. Behav., 7, 265–267.

Leon, M. and Moltz, H. (1972) *The development of the pheromonal bond in the albino rat.* Physiol. Behav., 8, 683–686.

Leshem, M. and Sherman, M. (2006) *Troubles shared are troubles halved: stress in rats is reduced in proportion to social propinquity.* Physiol. Behav., 89, 399–401.

Levai, O., Feistel, T., Breer, H. and Strotmann, J. (2006) *Cells in the vomeronasal organ express odorant receptors but project to the accessory olfactory bulb.* J. Comp. Neurol., 498, 476–490.

Levine, S. (1957) *Infantile experience and resistance to physiological stress.* Science, 126, 795–796.

Levine, S. and Broadhurst, P.L. (1963) *Genetic and ontogenetic determinants of adult behavior in the rat.* J. Comp. Physiol. Psychol., 56, 423–428.

Leyden, J.J., McGinley, K.J., Holzle, E., Labows, J.N. and Kligman, A.M. (1981) *The microbiology of the human axilla and its relationship to axillary odor.* J. Invest. Dermatol., 77, 413–416.

Leypold, B.G., Yu, C.R., Leinders-Zufall, T., Kim, M.M., Zufall, F. and Axel, R. (2002) *Altered sexual and social behaviors in trp2 mutant mice.* Proc. Nat. Acad. Sci. USA, 99, 6376–6381.

Li, W., Howard, J.D., Parrish, T.B. and Gottfried, J.A. (2008) *Aversive learning enhances perceptual and cortical discrimination of indiscriminable odor cues.* Science, 319, 1842–1845.

Liman, E.R., Corey, D.P. and Dulac, C. (1999) *TRP2: a candidate transduction channel for mammalian pheromone sensory signaling.* Proc. Nat. Acad. Sci. USA, 96, 5791–5796.

Little, B.B., Guzick, D.S., Malina, R.M. and Rocha Ferreira, M.D. (1989) *Environmental influences cause menstrual synchrony, not pheromones.* Am. J. Human Biol., 1, 53–57.

Lloyd-Thomas, A. and Keverne, E.B. (1982) *Role of the brain and accessory olfactory system in the block to pregnancy in mice.* Neurosci., 7, 907–913.

Lorenz, K. (1970) *The establishment of the instinct concept (1937).* In Studies in animal and human behaviour. Volume 1. Harvard University Press, Cambridge, MA, pp. 259–315.

Lott, D.F. and Hopwood, J.H. (1972) *Olfactory pregnancy-block in mice* (Mus musculus)*: an unusual response acquisition paradigm.* Anim. Behav., 20, 263–267.

Luderer, U., Morgan, M.S., Brodkin, C.A., Kalman, D.A. and Faustman, E.M. (1999) *Reproductive endocrine effects of acute exposure to toluene in men and women.* Occup. Environ. Med., 56, 657–666.

Ludvigson, H.W. (1969) *Runway behavior of the rat as a function of intersubject reward contingencies and constancy of daily reward schedule.* Psychonomic Sci., 15, 41–43.

Ludvigson, H.W., Mathis, D.A. and Choquette, K.A. (1985) *Different odors in rats from large and small rewards*. Anim. Learn. Behav., 13, 315–320.

Ludvigson, H.W. and Rottman, T.R. (1989) *Effects of ambient odors of lavender and cloves on cognition, memory, affect and mood*. Chem. Senses, 14, 525–536.

Lundström, J.N. (2005) Human pheromones: psychological and neurological modulation of a putative human pheromone. Uppsala Universitet, Uppsala, Sweden, pp. 1–73.

Lundström, J.N. and Olsson, M.J. (2005) *Subthreshold amounts of social odorant affect mood, but not behavior, in heterosexual women when tested by a male, but not a female, experimenter*. Biol. Psychol., 70, 197–204.

Luo, M., Fee, M.S. and Katz, L.C. (2003) *Encoding pheromonal signals in the accessory olfactory bulb of behaving mice*. Science, 299, 1196–1201.

Luque-Larena, J.J., López, P. and Gosálbez, J. (2001) *Scent matching modulates space use and agonistic behaviour between male snow voles*, Chionomys nivalis. Anim. Behav., 62, 1089–1095.

Lydell, K. and Doty, R.L. (1972) *Male rat of odor preferences for female urine as a function of sexual experience, urine age, and urine source*. Horm. Behav., 3, 205–212.

Ma, W., Miao, Z. and Novotny, M.V. (1998) *Role of the adrenal gland and adrenal-mediated chemosignals in suppression of estrus in the house mouse: the Lee-Boot effect revisited*. Biol. Reprod., 59, 1317–1320.

Ma, W., Miao, Z. and Novotny, M.V. (1999) *Induction of estrus in grouped female mice* (Mus domesticus) *by synthetic analogues of preputial gland constituents*. Chem. Senses, 24, 289–293.

Maarse, H. (1991) Volatile compounds in foods and beverages. Marcel Dekker, New York.

Mackay-Sim, A. and Laing, D.G. (1980) *Discrimination of odors from stressed rats by non-stressed rats*. Physiol. Behav., 24, 699–704.

Mackay-Sim, A. and Laing, D.G. (1981) *Rats' responses to blood and body odors of stressed and non-stressed conspecifics*. Physiol. Behav., 27, 503–510.

Mackintosh, J.H. and Grant, E.C. (1966) *The effect of olfactory stimuli on the agonistic behaviour of laboratory mice*. Z. Tierpsychol., 23, 584–587.

MacNiven, E., deCatanzaro, D. and Younglai, E.V. (1992) *Chronic stress increases estrogen and other steroids in inseminated rats*. Physiol. Behav., 52, 159–162.

Macrides, F., Bartke, A., Fernandez, F. and D'Angelo, W. (1974) *Effects of exposure to vaginal odor and receptive females on plasma testosterone in the male hamster*. Neuroendocrinology, 15, 355–364.

Macrides, F., Clancy, A.N., Singer, A.G. and Agosta, W.C. (1984) *Male hamster investigatory and copulatory responses to vaginal discharge: an attempt to impart sexual significance to an arbitrary chemosensory stimulus.* Physiol. Behav., 33, 627–632.

Macrides, F., Johnson, P.A. and Schneider, S.P. (1977) *Responses of the male golden hamster to vaginal secretion and dimethyl disulfide: attraction versus sexual behavior.* Behav. Biol., 20, 377–386.

Maggio, J.C., Maggio, J.H. and Whitney, G. (1983) *Experience-based vocalization of male mice to female chemosignals.* Physiol. Behav., 31, 269–272.

Maier, I. and Müller, D.G. (1987) *Sexual pheromones in algae.* Biol. Bull., 170, 175.

Mainardi, D. (1963) *Elimanazione della barriera etologica all'Isolamento riproduttivo tra Mus musculus domesticus e M.m. bactrianus mediante azione sull'apprendimento infantile.* Inst. Lombardo—Acad. Sci. Lett., 97, 291–299.

Mainardi, D., Marsan, M. and Pasquali, A. (1965) *Causation of sexual preferences of the house mouse. The behaviour of mice reared by parents whose odour was artificially altered.* Atti. Soc. Ital. Sci. Nat., 104, 325–338.

Mandl, A.M. and Zuckerman, S. (1952) *Factors influencing the onset of puberty in albino rats.* J. Endocrinol., 8, 357–364.

Mar, A., Spreekmeester, E. and Rochford, J. (2000) *Antidepressants preferentially enhance habituation to novelty in the olfactory bulbectomized rat.* Psychopharmacology, 150, 52–60.

Marchlewska-Koj, A. (1983) *Pregnancy blocking by pheromones.* In Vandenbergh, J.G. (ed.), Pheromones and reproduction in mammals. Academic Press, New York, pp. 151–173.

Marchlewska-Koj, A. (1997) *Sociogenic stress and rodent reproduction.* Neurosci. Biobehav. Rev., 21, 699–703.

Marchlewska-Koj, A., Pochron, E., Galewicz-Sojecka, A. and Galas, J. (1994) *Suppression of estrus in female mice by the presence of conspecifics or by foot shock.* Physiol. Behav., 55, 317–321.

Marchlewska-Koj, A. and Zacharczuk-Kakietek, M. (1990) *Acute increase in plasma corticosterone level in female mice evoked by pheromones.* Physiol. Behav., 48, 577–580.

Marr, J.N. and Gardner, L.E., Jr. (1965) *Early olfactory experience and later social behavior in the rat: preference, sexual responsiveness, and care of young.* J. Genet. Psychol., 107, 167–174.

Marr, J.N. and Lilliston, L.G. (1969) *Social attachment in rats by odor and age.* Behaviour, 23, 277–282.

Marsden, H.M. and Bronson, F.H. (1965) *Strange male block to pregnancy: its absence in inbred mouse strains.* Nature, 207, 878.

Marshall, D.A., Doty, R.L., Lucero, D.P. and Slotnick, B.M. (1981) *Odor detection thresholds in the rat for the vapors of three related perfluorocarbons and ethylene glycol dinitrate.* Chem. Senses, 6, 421–433.

Martin, I.G. (1980) *"Homeochemic" intraspecific chemical signal.* J. Chem. Ecol., 6, 517–519.

Martinez-Ricos, J., Agustin-Pavon, C., Lanuza, E. and Martinez-Garcia, F. (2007) *Intraspecific communication through chemical signals in female mice: reinforcing properties of involatile male sexual pheromones.* Chem. Senses, 32, 139–148.

Maruniak, J.A., Desjardins, C. and Bronson, F.H. (1977) *Dominant-subordinate relationships in castrated male mice bearing testosterone implants.* Amer. J. Physiol., 233, E495–E499.

Mateo, J.M. and Johnston, R.E. (2000) *Retention of social recognition after hibernation in Belding's ground squirrels.* Anim. Behav., 59, 491–499.

Matsumoto-Oda, A., Hamai, M., Hayaki, H., Hosaka, K., Hunt, K.D., Kasuya, E., Kawanaka, K., Mitani, J.C., Takasaki, H. and Takahata, Y. (2007) *Estrus cycle asynchrony in wild female chimpanzees,* Pan troglodytes schweinfurthii. Behav. Ecol. Sociobiol., 61, 661–668.

Matteo, S. (1987) *The effect of job stress and job interdependency on menstrual cycle length, regularity and synchrony.* Psychoneuroendocrinology, 12, 467–476.

Mayr, E. (1974) *Behavior programs and evolutionary strategies.* Amer. Scientist, 62, 650–659.

Mazur, A., Booth, A. and Dabbs, J.M. Jr. (1992) *Testosterone and chess competition.* Social Psychol. Quart., 55, 70–77.

McCarty, R. and Southwick, C.H. (1977) *Cross-species fostering: effects on the olfactory preference of* Onychomys torridus *and* Peromyscus leucopus. Behav. Biol., 19, 255–260.

McCaul, K.D., Gladue, B.A. and Joppa, M. (1992) *Winning, losing, mood, and testosterone.* Horm. Behav., 26, 488–504.

McClintock, M.K. (1971) *Menstrual synchrony and suppression.* Nature, 229, 244–245.

McClintock, M.K. (1978) *Estrous synchrony and its mediation by airborne chemical communication* (Rattus norvegicus). Horm. Behav., 10, 264–275.

McClintock, M.K. (1981) *Social control of the ovarian cycle and the function of estrous synchrony.* Amer. Zoologist, 21, 243–256.

McClintock, M.K. (1984) *Estrous synchrony: modulation of ovarian cycle length by female pheromones.* Physiol. Behav., 32, 701–705.

McClure, T.J. (1959) *Temporary nutritional stress and infertility in mice.* Nature, 181, 1132.

McCoy, N.L. and Pitino, L. (2002) *Pheromonal influences on sociosexual behavior in young women.* Physiol. Behav., 75, 367–375.

McDougall, W. (1921) An introduction to social psychology. Luce, Boston.

McNair, D.M., Loor, M. and Droppleman, L. (2003) Profile of Mood States. Educational and Industrial Testing Service, San Diego, CA.

McNeilly, A.S., Cooper, K.J. and Crighton, D.B. (1970) *Modification of the oestrous cycle of the underfed rat induced by proximity of a male.* J. Reprod. Fertil., 22, 359-.

Meisami, E. and Bhatnagar, K.P. (1998) *Structure and diversity in mammalian accessory olfactory bulb.* Microsc. Res. Tech., 43, 476–499.

Meisami, E., Mikhail, L., Baim, D. and Bhatnagar, K.P. (1998) *Human olfactory bulb: aging of glomeruli and mitral cells and a search for the accessory olfactory bulb.* Ann. NY Acad. Sci., 855, 708–715.

Melrose, D.R., Reed, H.C. and Patterson, R.L. (1971) *Androgen steroids associated with boar odour as an aid to the detection of oestrus in pig artificial insemination.* Brit. Vet. J., 127, 497–502.

Mennella, J.A. and Beauchamp, G.K. (1991a) *Maternal diet alters the sensory qualities of human milk and the nursling's behavior.* Pediatrics, 88, 737–744.

Mennella, J.A. and Beauchamp, G.K. (1991b) *The transfer of alcohol to human milk. Effects on flavor and the infant's behavior.* New Engl. J. Med., 325, 981–985.

Mennella, J.A. and Beauchamp, G.K. (1996) *The human infants' response to vanilla flavors in mother's milk and formula.* Infant Beh. Dev., 19, 13–19.

Mennella, J.A., Johnson, A. and Beauchamp, G.K. (1995) *Garlic ingestion by pregnant women alters the odor of amniotic fluid.* Chem. Senses, 20, 207–209.

Meredith, M. (1980) *The vomeronasal organ and accessory olfactory system in the hamster.* In Müller-Schwarze, D. and Silverstein, R.M. (eds.), Chemical signals. Plenum, New York, pp. 303–326.

Meredith, M. (1998) *Vomeronasal, olfactory, hormonal convergence in the brain. Cooperation or coincidence?* Ann. NY Acad. Sci., 855, 349–361.

Meredith, M. (2001) *Human vomeronasal organ function: a critical review of best and worst cases.* Chem. Senses, 26, 433–445.

Meredith, M., Marques, D.M., O'Connell, R.O. and Stern, F.L. (1980) *Vomeronasal pump: significance for male hamster sexual behavior.* Science, 207, 1224–1226.

Michael, R.P. and Keverne, E.B. (1968) *Pheromones in the communication of sexual status in primates.* Nature, 218, 746–749.

Michael, R.P. and Keverne, E.B. (1970a) *Primate sex pheromones of vaginal origin.* Nature, 225, 84–85.

Michael, R.P. and Keverne, E.B. (1970b) *A male sex-attractant pheromone in rhesus monkey vaginal secretions.* J. Endocrinol., 46, xx–xxi.

Michael, R.P., Keverne, E.B. and Bonsall, R.W. (1971) *Pheromones: isolation of male sex attractants from a female primate.* Science, 172, 964–966.

Michael, R.P., Zumpe, D., Keverne, E.B. and Bonsall, R.W. (1972) *Neuroendocrine factors in the control of primate behavior.* Recent Prog. Horm. Res., 28, 665–706.

Milenkovic, L., Bogic, L. and Martinovic, J.V. (1986) *Effects of oestradiol and progesterone on stress-induced secretion of prolactin in ovariectomized and/ or adrenalectomized female rats.* Acta Endocrinologica, 112, 79–82.

Mody, J.K. and Christian, J.J. (1962) *Adrenals and reproductive organs of female mice kept singly, grouped or grouped with a vasectomized male.* J. Endocrinol., 24, 1–6.

Mogg, K., Bradley, B.P., Williams, R. and Mathews, A. (1993) *Subliminal processing of emotional information in anxiety and depression.* J. Abnorm. Psychol., 102, 304–311.

Moltz, H. and Leidahl, L.C. (1977) *Bile, prolactin, and the maternal pheromone.* Science, 196, 81–83.

Mombaerts, P., Wang, F., Dulac, C., Chao, S.K., Nemes, A., Mendelsohn, M., Edmondson, J. and Axel, R. (1996) *Visualizing an olfactory sensory map.* Cell, 87, 675–686.

Moncho-Bogani, J., Lanuza, E., Hernandez, A., Novejarque, A. and Martinez-Garcia, F. (2002) *Attractive properties of sexual pheromones in mice: innate or learned?* Physiol. Behav., 77, 167–176.

Moncho-Bogani, J., Lanuza, E., Lorente, M.J. and Martinez-Garcia, F. (2004) *Attraction to male pheromones and sexual behaviour show different regulatory mechanisms in female mice.* Physiol. Behav., 81, 427–434.

Moncho-Bogani, J., Martinez-Garcia, F., Novejarque, A. and Lanuza, E. (2005) *Attraction to sexual pheromones and associated odorants in female mice involves activation of the reward system and basolateral amygdala.* Eur. J. Neurosci., 21, 2186–2198.

Monder, H., Lee, C.T., Donovick, P.J. and Burright, R. (1978) *Male mouse urine extract effects on pheromonally mediated reproductive functions of female mice.* Physiol. Behav., 20, 447–452.

Monti-Bloch, L., Diaz-Sanchez, V., Jennings-White, C. and Berliner, D.L. (1998) *Modulation of serum testosterone and autonomic function through stimulation of the male human vomeronasal organ (VNO) with pregna-4,20-diene-3,6-dione.* J. Steroid Biochem. Mol. Biol., 65, 237–242.

Monti-Bloch, L. and Grosser, B.I. (1991) *Effect of putative pheromones on the electrical activity of the human vomeronasal organ and olfactory epithelium.* J. Steroid Biochem. Mol. Biol., 39, 573–582.

Monti-Bloch, L., Jennings-White, C., Dolberg, D.S. and Berliner, D.L. (1994) *The human vomeronasal system.* Psychoneuroendocrinology, 19, 673–686.

Montigny, D., Coureaud, G. and Schaal, B. (2006) *Rabbit pup response to the mammary pheromone: from automatism to prandial control.* Physiol. Behav., 89, 742–749.

Moore, J.E., Pelosi, P. and Forrester, I.J. (1977) *Specific anosmia to 5α-androst-16en-3-one and gamma-pentadecalactone: the urinous and musky primary odors.* Chem. Senses Flav., 2, 401–425.

Moore, R.E. (1965) *Olfactory discrimination as an isolating mechanism between* Peromyscus maniculatus *and* Peromyscus polionotus. Am. Midl. Nat., 73, 85–100.

Moran, D.T., Jafek, B.W. and Rowley, I.J.C. (1991) *The vomeronasal (Jacobson's) organ in man: ultrastructure and frequency of occurrence.* J. Steroid Biochem. Mol. Biol., 39, 545–552.

Moran, D.T., Monti-Bloch, L., Stensaas, L.J. and Berliner, D.L. (1995) *Structure and function of the human vomeronasal organ.* In Doty, R.L. (ed.), Handbook of olfaction and gustation. Marcel Dekker, Inc., New York, pp. 793–820.

Moriceau, S. and Sullivan, R.M. (2004) *Unique neural circuitry for neonatal olfactory learning.* J. Neurosci., 24, 1182–1189.

Moriceau, S. and Sullivan, R.M. (2006) *Maternal presence serves as a switch between learning fear and attraction in infancy.* Nat. Neurosci., 9, 1004–1006.

Morris, N.M. and Udry, J.R. (1978) *Pheromonal influences on human sexual behavior: an experimental search.* J. Biosoc. Sci., 10, 147–157.

Morton, J.R.C., Denenberg, V.H. and Zarrow, M.X. (1963) *Modification of sexual development through stimulation in infancy.* Endocrinology, 72, 439–442.

Moss, M., Hewitt, S., Moss, L. and Wesnes, K. (2008) *Modulation of cognitive performance and mood by aromas of peppermint and ylang-ylang.* Int. J. Neurosci., 118, 59–77.

Mossman, C.A. and Drickamer, L.C. (1996) *Odor preferences of female house mice* (Mus domesticus) *in seminatural enclosures.* J. Comp. Psychol., 110, 131–138.

Mouly, A.M., Fort, A., Ben-Boutayab, N. and Gervais, R. (2001) *Olfactory learning induces differential long-lasting changes in rat central olfactory pathways.* Neurosci., 102, 11–21.

Mucignat-Caretta, C. (2002) *Modulation of exploratory behavior in female mice by protein-borne male urinary molecules.* J. Chem. Ecol., 28, 1853–1863.

Mucignat-Caretta, C., Caretta, A. and Cavaggioni, A. (1995a) *Acceleration of puberty onset in female mice by male urinary proteins.* J. Physiol., 486, 517–522.

Mucignat-Caretta, C., Caretta, A. and Cavaggioni, A. (1995b) *Pheromonally accelerated puberty is enhanced by previous experience of the same stimulus.* Physiol. Behav., 57, 901–903.

Mucignat-Caretta, C., Cavaggioni, A. and Caretta, A. (2004) *Male urinary chemosignals differentially affect aggressive behavior in male mice.* J. Chem. Ecol., 30, 777–791.

Mugford, R.A. and Nowell, N.W. (1970) *Pheromones and their effect on aggression in mice.* Nature, 226, 967–968.

Mugford, R.A. and Nowell, N.W. (1971) *Shock-induced release of the preputial gland secretions that elicit fighting in mice.* J. Endocrinol., 51, xvi–xvii.

Mugford, R.A. and Nowell, N.W. (1972) *The dose-response to testosterone propionate of preputial glands, pheromones and aggression in mice.* Horm. Behav., 3, 39–46.

Mukae, H. and Sato, M. (1992) *The effect of color temperature of lighting sources on the autonomic nervous functions.* Ann. Physiol. Anthropol., 11, 533–538.

Müller-Schwarze, D. (1971) *Pheromones in black-tailed deer* (Odocoileus hemionus columbianus). Anim. Behav., 19, 141–152.

Müller-Schwarze, D. (1977) *Complex mammalian behavior and pheromone bioassay in the field.* In Müller-Schwarze, D. and Mozell, M.M. (eds.), Chemical signals in vertebrates. Plenum Press, New York, pp. 413–433.

Müller-Schwarze, D., Altieri, R. and Porter, N. (1998) *Alert odor from skin gland in deer.* J. Chem. Ecol., 10, 1707–1729.

Müller-Schwarze, R., Müller-Schwarze, D., Singer, A.G. and Silverstein, R.M. (1974) *Mammalian pheromone: identification of active component in the subauricular scent of the male pronghorn.* Science, 183, 860–862.

Müller-Velten, H. (2003) *Uber den Angstgeruch bei der Hausmaus* (Mus musculus L). Z. Vergleich. Physiol., 52, 401–429.

Murphy, M.R. (1980) *Sexual preferences of male hamsters: importance of preweaning and adult experience, vaginal secretion, and olfactory or vomeronasal sensation.* Behav. Neur. Biol., 30, 323–340.

Murphy, M.R. and Schneider, G.E. (1970) *Olfactory bulb removal eliminates mating behavior in the male golden hamster.* Science, 167, 302–304.

Mykytowycz, R. (1973) *Reproduction of mammals in relation to environmental odours.* J. Repr. Fertil., Suppl. 19, 433–446.

Mykytowycz, R. and Dudzinski, M.L. (1966) *A study of the weight of odoriferous and other glands in relation to social status and degree of sexual activity in the wild rabbit,* Oryctolagus cuniculus (L.). CSIRO Wildlife Research, 11, 31–47.

Nagai, M., Wada, M., Usui, N., Tanaka, A. and Hasebe, Y. (2000) *Pleasant odors attenuate the blood pressure increase during rhythmic handgrip in humans.* Neurosci. Lett., 289, 227–229.

Nakagawa, R., Tanaka, M., Kohno, Y., Noda, Y. and Nagasaki, N. (1981) *Regional responses of rat brain noradrenergic neurons to acute intense stress.* Pharmacol. Biochem. Behav., 14, 729–732.

Nakamura, T., Tanida, M., Niijima, A., Hibino, H., Shen, J. and Nagai, K. (2007) *Auditory stimulation affects renal sympathetic nerve activity and blood pressure in rats.* Neurosci. Lett., 416, 107–112.

Nakata, K., Tsuji, K., Holldobler, B. and Taki, A. (1998) *Sexual calling by workers using the metatibial glands in the ant,* Diacamma *sp., from Japan (Hymenoptera: Formicidae).* J. Insect Behav., 11, 869–877.

Narendran, R., Etches, R.J., Hurnik, J.F. and Bowman, G.H. (1980) *Effect of social hierarchy on plasma 5-α-androstenone levels in boars.* Can. J. Anim. Sci., 60, 1061.

Nichols, D.J. and Chevins, P.F.D. (1981) *Effects of housing on corticosterone rhythm and stress responses in female mice.* Physiol. Behav., 27, 1–5.

Nicolaides, N. (1974) *Skin lipids: their biochemical uniqueness.* Science, 186, 19–26.

Nigrosh, B.J., Slotnick, B.M. and Nevin, J.A. (1975) *Olfactory discrimination, reversal learning, and stimulus control in rats.* J. Comp. Physiol. Psychol., 89, 285–294.

Niijima, A. and Nagai, K. (2003) *Effect of olfactory stimulation with flavor of grapefruit oil and lemon oil on the activity of sympathetic branch in the white adipose tissue of the epididymis.* Exp. Biol. Med., 228, 1190–1192.

Nishimura, K., Utsumi, K., Yuhara, M. Fujitani, Y. and Iritani, A. (1989) *Identification of puberty-accelerating pheromones in male mouse urine.* J. Exp. Zool., 251, 300–305.

Nixon, A., Mallet, A.I. and Gower, D.B. (1988) *Simultaneous quantification of five odorous steroids (16-androstenes) in the axillary hair of men.* J. Steroid Biochem., 29, 505–510.

Nodari, F., Hsu, F.F., Fu, X., Holekamp, T.F., Kao, L.F., Turk, J. and Holy, T.E. (2008) *Sulfated steroids as natural ligands of mouse pheromone-sensing neurons.* J. Neurosci., 28, 6407–6418.

Noguchi, H. and Sakaguchi, T. (1999) *Effect of illuminance and color temperature on lowering of physiological activity.* Appl. Human Sci., 18, 117–123.

Norris, M.L. and Adams, C.E. (1979) *Exteroceptive factors and pregnancy block in the Mongolian gerbil,* Meriones unguiculatus. J. Reprod. Fertil., 57, 401–404.

Novotny, M., Harvey, S., Jemiolo, B. and Alberts, J. (1985) *Synthetic pheromones that promote inter-male aggression in mice.* Proc. Nat. Acad. Sci. USA, 82, 2059–2061.

Novotny, M., Jemiolo, B. and Harvey, S. (1990) *Chemistry of rodent pheromones: molecular insights into chemical signaling in mammals.* In MacDonald, D.W., Muller-Schwarze, D. and Natynczuk, S.E. (eds.), Chemical signals in vertebrates. Vol. 5. Oxford University Press, Oxford, pp. 1–22.

Novotny, M.V., Jemiolo, B., Wiesler, D., Ma, W., Harvey, S., Xu, F., Xie, T.M. and Carmack, M. (1999a) *A unique urinary constituent, 6-hydroxy-6-methyl-3-heptanone, is a pheromone that accelerates puberty in female mice.* Chem. Biol., 6, 377–383.

Novotny, M.V., Ma, W., Wiesler, D. and Zidek, L. (1999b) *Positive identification of the puberty-accelerating pheromone of the house mouse: the volatile ligands associating with the major urinary protein.* Proc. Roy. Soc. Lond. B: Biol. Sci., 266, 2017–2022.

Nowell, N.W. and Wouters, A. (1973) *Effect of cyproterone acetate upon aggressive behavior in the laboratory mouse.* J. Endocrinol., 57, R36–R37.

Nyakas, C. and Endröczi, E. (1970) *Olfaction guided approaching behaviour of infantile rats to the mother in maze box.* Acta Physiol. Academ. Sci. Hung., 38, 59–65.

Nyby, J. (2001) *Auditory communication among adults.* In Willott, J.F. (ed.), Handbook of mouse auditory research: from behavior to molecular biology. CRC Press, Boca Raton, pp. 3–18.

Nyby, J., Dizinno, G.A. and Whitney, G. (1976) *Social status and ultrasonic vocalizations of male mice.* Behav. Biol., 18, 285–289.

Nyby, J., Whitney, G., Schmitz, S. and Dizinno, G. (1978) *Postpubertal experience establishes signal value of mammalian sex odor.* Behav. Biol., 22, 545–552.

Nyby, J., Wysocki, C.J., Whitney, G. and Dizinno, G. (1977) *Pheromonal regulation of male mouse ultrasonic courtship* (Mus musculus). Animal Behaviour, 25, 333–341.

O'Connell, R.J., Singer, A.G., Macrides, F., Pfaffmann, C. and Agosta, W.C. (1978) *Responses of the male golden hamster to mixtures of odorants identified from vaginal discharge.* Behav. Biol., 24, 244–255.

Ohloff, G., Maurer, B., Winter, B. and Wolfgang, G. (1983) *Structural and configurational dependence of the sensory process in steroids.* Helv. Chir. Acta, 192–201.

Oinonen, K.A. and Mazmanian, D. (2002) *To what extent do oral contraceptives influence mood and affect?* J. Affect. Dis., 70, 229–240.

Olsson, M.J., Lundstrom, J.N., Diamantopoulou, S. and Esteves, F. (2006) *A putative female pheromone affects mood in men differently depending on social context.* Eur. Rev. Appl. Psychol.—Rev. Eur. Psychol. Appl., 56, 279–284.

Osgood, C.E., Suci, G.J. and Tannenbaum, P.H. (1957) The measurement of meaning. University of Illinois Press, Urbana.

Otto, T., Cousens, G. and Rajewski, K. (1997) *Odor-guided fear conditioning in rats: 1. Acquisition, retention, and latent inhibition.* Behav. Neurosci., 111, 1257–1264.

Over, R., Cohen-Tannoudji, J., Dehnhard, M., Claus, R. and Signoret, J.P. (1990) *Effect of pheromones from male goats on LH-secretion in anoestrous ewes.* Physiol. Behav., 48, 665–668.

Owen, K. and Thiessen, D.D. (1974) *Estrogen and progesterone interaction in the regulation of scent marking in the female Mongolian gerbil* (Meriones unguiculatus). Physiol. Behav., 12, 351–355.

Parkes, A.S. and Bruce, H.M. (1961) *Olfactory stimuli in mammalian reproduction.* Science, 134, 1049–1054.

Parkes, A.S. and Bruce, H.M. (1962) *Pregnancy block in female mice placed in boxes soiled by males.* J. Reprod. Fertil., 4, 303–308.

Patterson, R.L.S. (1966) *Possible contribution of phenolic components to boar odour.* Nature (London), 212, 744–745.

Patterson, R.L.S. (1968) *Acidic components of boar preputial fluid.* J. Sci. Food Agricult., 19, 38–40.

Paul, T., Schiffer, B., Zwarg, T., Krueger, T.H.C., Karama, S., Schedlowski, M., Forsting, M. and Gizewski, E.R. (2008) *Brain response to visual sexual stimuli in heterosexual and homosexual males.* Hum. Brain Mapp., 29, 726–735.

Pause, B.M. (2004) *Are androgen steroids acting as pheromones in humans?.* Physiol. Behav., 83, 21–29.

Pedersen, P.E. and Blass, E.M. (1982) *Prenatal and postnatal determinants of the 1st suckling episode in albino rats.* Dev. Psychobiol., 15, 349–355.

Pedersen, P.E., Stewart, W.B., Greer, C.A. and Shepherd, G.M. (1983) *Evidence for olfactory function in utero.* Science, 221, 478–480.

Penn, D. and Potts, W. (1998) *MHC-disassortative mating preferences reversed by cross-fostering.* Proc. Roy. Soc. Lond. B: Biol. Sci., 265, 1299–1306.

Pereira, M.E. (1991) *Asynchrony within estrous synchrony among ringtailed lemurs (Primates: Lemuridae).* Physiol. Behav., 49, 47–52.

Petit, C., Hossaert-McKey, M., Perret, P., Blondel, J., Lambright, C. and Lambrechts, M.M. (2002) *Blue tits use selected plants and olfaction to maintain an aromatic environment for nestlings.* Ecol. Lett., 5, 585–589.

Petrulis, A. and Johnston, R.E. (1995) *A reevaluation of dimethyl disulfide as a sex attractant in golden hamsters.* Physiol. Behav., 57, 779–784.

Pfaff, D.W., Phillips, I.M. and Rubin, R.T. (2004) Principles of hormone/behavior relations. Elsevier Academic Press, Burlington, MA.

Pihet, S., Mellier, D., Bullinger, A. and Schaal, B. (1997) *Réponses comportemen-tales aux odeurs chez le nouveau-né prématuré: étude préliminaire.* Enfance, 1, 33–46.

Pinker, S. (2007) The stuff of thought: language as a window into human nature. Penguin Group, New York.

Pohorecky, L.A., Blakley, G.G., Ma, E.W., Soini, H.A., Wiesler, D., Bruce, K.E. and Novotny, M.V. (2008) *Social housing influences the composition of volatile compounds in the preputial glands of male rats.* Horm. Behav., 53, 536–545.

Poindron, P., Nowak, R., Levy, F., Porter, R.H. and Schaal, B. (1993) *Development of exclusive mother-young bonding in sheep and goats.* Oxford Rev. Reprod. Biol., 15, 311–364.

Porter, R.H., Matochik, J.A. and Makin, J.W. (1983) *Evidence for phenotype matching in spiny mice* (Acomys cahirinus). Anim. Behav., 31, 978–984.

Porter, R.H. and Schaal, B. (2003) *Olfaction and the development of social behavior in neonatal mammals.* In Doty, R.L. (ed.), Handbook of olfaction and gustation. Marcel Dekker, New York, pp. 309–327.

Porter, R.H. and Winberg, J. (1999) *Unique salience of maternal breast odors for newborn infants.* Neurosci. Biobehav. Rev., 23, 439–449.

Potiquet, M. (1891) *Le canal Jacobson.* Revue Laryngol. Otol. Rhinol. (Bordeaux), 2, 737–753.

Powers, J.B., Fields, R.B. and Winans, S.S. (1979) *Olfactory and vomeronasal system participation in male hamsters' attraction to female vaginal secretions.* Physiol. Behav., 22, 77–84.

Powers, J.B. and Winans, S.S. (1973) *Sexual behavior in peripherally anosmic male hamsters.* Physiol. Behav., 10, 361–368.

Prelog, V. and Ruzicka, L. (1944) *Untersuchungen über Organextrakte. 5. Mitteilung: über zwei muschusartig riechende Steroide aus Schweines-testes-extracten.* Helv. Chir. Acta, 27, 61–66.

Prelog, V., Ruzicka, L., Meister, P. and Wieland, P. (1945) *Steroiden und Sexualhormonen. 113. Mitteilung: Untersuchungen über den Zusammenhang zwischen Konstitution und der Geruch bei Steroiden.* Helv. Chir. Acta, 28, 618–627.

Preti, G. (1987) *Reply to Wilson.* Horm. Behav., 21, 547–550.

Preti, G., Cutler, W.B., Garcia, C.R., Huggins, G.R. and Lawley, H.J. (1986) *Human axillary secretions influence women's menstrual cycles: the role of donor extract of females.* Horm. Behav., 20, 474–482.

Price, E., Dally, M., Erhard, H., Gerzevske, M., Kelly, M., Moore, N., Schultze, A. and Topper, C. (1998) *Manipulating odor cues facilitates add-on fostering in sheep.* J. Anim. Sci., 76, 961–964.

Price, E.O. (1975) *Hormonal control of urine-marking in wild and domestic Norway rats.* Horm. Behav., 6, 393–397.

Price, E.O., Dunn, G.C., Talbot, J.A. and Dally, M.R. (1984) *Fostering lambs by odor transfer: the substitution experiment.* J. Anim. Sci., 59, 301–307.

Price, M.A. and Vandenbergh, J.G. (1992) *Analysis of puberty-accelerating pheromones.* J. Exp. Zool., 264, 42–45.

Purvis, K., Cooper, K.J. and Haynes, N.B. (1971) *The influence of male proximity and dietary restriction on the oestrous cycle of the rat.* J. Reprod. Fertil., 27, 167–176.

Quadagno, D.M. and Banks, E.M. (1970) *The effects of reciprocal cross fostering on the behaviour of two species of rodents,* Mus musculus *and* Baiomys taylori ater. Anim. Behav., 18, 379–390.

Quadagno, D.M., Shubeita, H.E., Deck, J. and Francoer, D. (1981) *The effects of males, athletic activities and all female living conditions on the menstrual cycle.* Psychoneuroendocrinology, 6, 239–244.

Rajendren, G. and Dominic, C.J. (1985) *Effect of transection of the vomeronasal nerve on the male-induced implantation failure (the Bruce effect) in mice.* Indian J. Exp. Biol., 23, 635–637.

Rajendren, G. and Dominic, C.J. (1987) *The male-induced pregnancy block (the Bruce effect) in mice: re-evaluation of the ability of exogenous progesterone in preventing implantation failure.* Exp. Clin. Endocrinol., 89, 188–196.

Rajendren, G. and Dutta, A.K. (1988) *Effect of haloperidol on male-induced implantation failure (Bruce effect) in mice.* Indian J. Exp. Biol., 26, 909–910.

Rako, S. and Friebely, J. (2004) *Pheromonal influences on sociosexual behavior in postmenopausal women.* J. Sex. Res., 41, 372–380.

Ramaley, J.A. (1977) *Changes in adrenocortical function precede the onset of persistent estrus in rats exposed to constant light before puberty.* Biol. Reprod., 17, 733–737.

Ramm, S.A., Cheetham, S.A. and Hurst, J.L. (2008) *Encoding choosiness: female attraction requires prior physical contact with individual male scents in mice.* Proc. Soc. Exp. Biol. Med., 275, 1727–1735.

Rampin, O., Jérôme, N., Briant, C., Boué, F. and Maurin, Y. (2006) *Are oestrus odours species specific?* Behav. Brain Res., 172, 169–172.

Rasmussen, J.L., Rajecki, D.W. and Craft, H.D. (1993) *Humans' perceptions of animal mentality: ascriptions of thinking.* J. Comp. Psychol., 107, 283–290.

Rasmussen, L.E., Lee, T.D., Roelofs, W.L., Zhang, A. and Daves, G.D., Jr. (1996) *Insect pheromone in elephants.* Nature, 379, 684.

Rasmussen, L.E., Lee, T.D., Zhang, A., Roelofs, W.L. and Daves, G.D., Jr. (1997) *Purification, identification, concentration and bioactivity of (Z)-7-dodecen-*

1-yl acetate: sex pheromone of the female Asian elephant, Elephas maximus. Chem. Senses, 22, 417–437.

Rasmussen, L.E.L. (1998) *Chemical communication: an integral part of functional Asian elephant* (Elephas maximus) *society.* Ecoscience, 5, 410–426.

Rasmussen, L.E.L. and Greenwood, D.R. (2003) *Frontalin: a chemical message of musth in Asian elephants* (Elephas maximus). Chem. Senses, 28, 433–446.

Rasmussen, L.E.L. and Krishnamurthy, V. (2000) *How chemical signals integrate Asian elephant society: the known and the unknown.* Zoo Biology, 19, 405–423.

Rasmussen, L.E.L., Riddle, H.S. and Krishnamurthy, V. (2002) *Mellifluous matures to malodorous in musth.* Nature, 415, 975–976.

Rasmussen, L.E.L. and Schulte, B.A. (1998) *Chemical signals in the reproduction of Asian* (Elephas maximus) *and African* (Loxodonta africana) *elephants.* Anim. Reprod. Sci., 53, 19–34.

Reed, H.C., Melrose, D.R. and Patterson, R.L. (1974) *Androgen steroids as an aid to the detection of oestrus in pig artificial insemination.* Brit. Vet. J., 130, 61–67.

Reid, I.C. and Morris, R.G. (1993) *The enigma of olfactory learning.* Trend Neurosci., 16, 17–20.

Rejeski, W.J., Gagne, M., Parker, P.E. and Koritnik, D.R. (1989) *Acute stress reactivity from contested dominance in dominant and submissive males.* Behav. Med., 15, 118–124.

Reynierse, J.H. (1974) *Communication elements constraining animal learning and performance.* In Krames, L., Pliner, P. and Allow, T. (eds.), Nonverbal communication. Plenum, New York, pp. 1–24.

Reynolds, J. and Keverne, E.B. (1979) *The accessory olfactory system and its role in the pheromonally mediated suppression of oestrus in grouped mice.* J. Reprod. Fertil., 57, 31–35.

Rich, T.J. and Hurst, J.L. (1999) *The competing countermarks hypothesis: reliable assessment of competitive ability by potential mates.* Anim. Behav., 58, 1027–1037.

Richards, D.B. and Stevens, D.A. (1974) *Evidence for marking with urine by rats.* Behav. Biol., 12, 517–523.

Richardson, R. and McNally, G.P. (2003) *Effects of an odor paired with illness on startle, freezing, and analgesia in rats.* Physiol. Behav., 78, 213–219.

Robin, O., Alaoui-Ismaili, O., Dittmar, A. and Vernet-Maury, E. (1999) *Basic emotions evoked by eugenol odor differ according to the dental experience. A neurovegetative analysis.* Chem. Senses, 24, 327–335.

Rodriguez Echandia, E.L., Foscolo, M. and Broitman, S.T. (1982) *Preferential nesting in lemon-scented environment in rats reared on lemon-scented bedding from birth to weaning.* Physiol. Behav., 29, 47–49.

Rodriguez, I., Greer, C.A., Mok, M.Y. and Mombaerts, P. (2000) *A putative pheromone receptor gene expressed in human olfactory mucosa.* Nat. Genet., 26, 18–19.

Rogers, J.G., Jr. and Beauchamp, G.K. (1976) *Some implications of primer chemical stimuli in rodents.* In Doty, R.L. (ed.), Mammalian olfaction, reproductive processes, and behavior. Academic Press, New York, pp. 181–197.

Rolls, E.T. (2005) *Taste, olfactory, and food texture processing in the brain, and the control of food intake.* Physiol. Behav., 85, 45–56.

Romantshik, Q., Porter, R.H., Tillman, V., and Varendi, H. (2007) *Preliminary evidence of a sensitive period of olfactory learning by human newborns.* Acta Paediatr., 96, 372–376.

Romeo, R.D., Parfitt, D.B., Richardson, H.N. and Sisk, C.L. (1998) *Pheromones elicit equivalent levels of Fos-immunoreactivity in prepubertal and adult male Syrian hamsters.* Horm. Behav., 34, 48–55.

Ropartz, P. (1966) *Mise en evidence d'une odeur de groupe chez les souris par a la mesure de l'activite locomotrice.* C. R. Acad. Sci. Biol., 262, 507–510.

Rose, E. and Drickamer, L.C. (1975) *Castration, sexual experience, and female urine odor preferences in adult BDF1 male mice.* Bull. Psychonom. Soc., 5, 84–86.

Rosell, F. and Sanda, J. (2006) *Potential risks of olfactory signaling: the effect of predators on scent marking by beavers.* Beh. Ecol., 17, 897–904.

Rosser, A.E. and Keverne, E.B. (1985) *The importance of central noradrenergic neurones in the formation of an olfactory memory in the prevention of pregnancy block.* Neurosci., 15, 1141–1147.

Roth, T.L., Moriceau, S. and Sullivan, R.M. (2006) *Opioid modulation of Fos protein expression and olfactory circuitry plays a pivotal role in what neonates remember.* Learn. Mem., 13, 590–598.

Rottman, S.J. and Snowdon, C.T. (1972) *Demonstration and analysis of an alarm pheromone in mice.* J. Comp. Physiol. Psychol., 81, 483–490.

Runner, M.N. (1959) *Embryocidal effect of handling pregnant mice and its prevention with progesterone.* Anat. Rec., 133, 330–331.

Rusak, B. and Zucker, I. (1979) *Neural regulation of circadian rhythms.* Physiol. Rev., 59, 449–526.

Russell, M.J. (1976) *Human olfactory communication.* Nature, 260, 520–522.

Russell, M.J., Switz, G.M. and Thompson, K. (1980) *Olfactory influences on the human menstrual cycle.* Pharm. Biochem. Behav., 13, 737–738.

Rutowski, R.L. (1981) *The function of pheromones.* J. Chem. Ecol., 7, 481–484.

Ryan, K.D. and Schwartz, N.B. (1977) *Grouped female mice: demonstration of pseudopregnancy.* Biol. Reprod., 17, 583.

Ryon, J., Fentress, J.C., Harrington, F.H. and Bragdon, S. (1986) *Scent rubbing in wolves* (Canis lupus): *the effect of novelty.* Can. J. Zool., 64, 573–577.

Sachs, B.D. (1996) *Penile erection in response to remote cues from females: albino rats severely impaired relative to pigmented strains.* Physiol. Behav., 60, 803–808.

Sachs, B.D. (1997) *Erection evoked in male rats by airborne scent from estrous females.* Physiol. Behav., 62, 921–924.

Sachs, B.D., Akasofu, K., Citron, J.H., Daniels, S.B. and Natoli, J.H. (1994) *Noncontact stimulation from estrous females evokes penile erection in rats.* Physiol. Behav., 55, 1073–1079.

Sackler, A.M., Weldman, A.S., Bradshaw, M. and Jurtshuk, P. (1959) *Endocrine changes due to auditory stress.* Acta Endocrinol., 31, 405–418.

Sahu, S.C. and Dominic, C.J. (1980) *Chlorpromazine inhibition of the pheromonal block to pregnancy (the Bruce effect) in mice.* Indian J. Exp. Biol., 18, 1025–1027.

Sahu, S.C. and Dominic, C.J. (1981) *Failure of adrenalectomy to prevent the male-induced pregnancy block (Bruce effect) in mice.* Endokrinologie, 78, 156–160.

Sahu, S.C. and Dominic, C.J. (1983) *Masking effect of artificial scents on pheromonal block to pregnancy (the Bruce effect) in mice.* Indian J. Exp. Biol., 21, 497–499.

Sales, G.D. (1972) *Ultrasound and mating behaviour in rodents with some observations on other behavioural situations.* J. Zool., 168, 149–164.

Sales, G.D. and Pye, D. (1974) Ultrasonic communication in mammals. Chapman and Hall, London.

Saletu, B., Saletu, M., Herrmann, W.M. and Itil, T.M. (1975) *Are hormones psychoactive? Evoked potential investigations in man.* Arzneimittel-Forschung, 25, 1321–1327.

Sam, M., Vora, S., Malnic, B., Ma, W., Novotny, M.V. and Buck, L.B. (2001) *Neuropharmacology. Odorants may arouse instinctive behaviours.* Nature, 412, 142.

Sarkamo, T., Tervaniemi, M., Laitinen, S., Forsblom, A., Soinila, S., Mikkonen, M., Autti, T., Silvennoinen, H.M., Erkkila, J., Laine, M., Peretz, I. and Hietanen, M. (2008) *Music listening enhances cognitive recovery and mood after middle cerebral artery stroke.* Brain, 131, 866–876.

Sarnat, H.B. (1978) *Olfactory reflexes in the newborn infant.* J. Pediat., 92, 624–626.

Savic, I. (2002) *Imaging of brain activation by odorants in humans.* Curr. Opinion Neurobiol., 12, 455–461.

Savic, I., Berglund, H., Gulyas, B. and Roland, P. (2001) *Smelling of odorous sex hormone-like compounds causes sex-differentiated hypothalamic activations in humans.* Neuron, 31, 661–668.

Savic, I., Berglund, H. and Lindstrom, P. (2005) *Brain response to putative pheromones in homosexual men.* Proc. Nat. Acad. Sci. USA, 102, 7356–7361.

Schaal, B., Coureaud, G., Langlois, D., Ginies, C., Semon, E. and Perrier, G. (2003) *Chemical and behavioural characterization of the rabbit mammary pheromone.* Nature, 424, 68–72.

Schaal, B., Marlier, L. and Soussignan, R. (1998) *Olfactory function in the human fetus: evidence from selective neonatal responsiveness to the odor of amniotic fluid.* Behav. Neurosci., 112, 1438–1449.

Schaal, B., Marlier, L. and Soussignan, R. (2000) *Human foetuses learn odours from their pregnant mother's diet.* Chem. Senses, 25, 729–737.

Schaal, B., Motagner, H., Hertling, E., Bolzoni, D., Moyse, R. and Quinchon, R. (1980) *Les stimulations olfactives dans les relations entre l'enfant et la mere.* Reprod. Nutr. Dev., 20, 843–858.

Schaal, B., Orgeur, P. and Arnould, C. (1995) *Olfactory preferences in newborn lambs: possible influences of prenatal experience.* Behaviour, 132, 351–365.

Schaal, B., Orgeur, P. and Marler, M. (1994) *Amniotic fluid odor in neonatal adaptation: a summary of recent research in mammals.* Adv. Biosci., 93, 239–245.

Schank, J.C. (1997) *Problems with dimensionless measurement models of synchrony in biological systems.* Am. J. Primatol., 41, 65–85.

Schank, J.C. (2000a) *Can pseudo entrainment explain the synchrony of estrous cycles among golden hamsters* (Mesocricetus auratus)? Horm. Behav., 38, 94–101.

Schank, J.C. (2000b) *Menstrual-cycle variability and measurement: further cause for doubt.* Psychoneuroendocrinology, 25, 837–847.

Schank, J.C. (2001a) *Measurement and cycle variability: reexamining the case for ovarian-cycle synchrony in primates.* Behav. Processes, 56, 131–146.

Schank, J.C. (2001b) *Menstrual-cycle synchrony: problems and new directions for research.* J. Comp. Psychol., 115, 3–15.

Schank, J.C. (2001c) *Oestrous and birth synchrony in Norway rats,* Rattus norvegicus. Anim. Behav., 62, 409–415.

Schank, J.C. (2001d) *Do Norway rats* (Rattus norvegicus) *synchronize their estrous cycles?* Physiol. Behav., 72, 129–139.

Schank, J.C. (2002) *A multitude of errors in menstrual-synchrony research: replies to Weller and Weller (2002) and Graham (2002).* J. Comp. Psychol., 116, 319–322.

Schank, J.C. (2006) *Do human menstrual-cycle pheromones exist?* Hum. Nature, 17, 448–470.

Schank, J.C. and Alberts, J.R. (2000) *Effects of male rat urine on reproductive and developmental parameters in the dam and her female offspring.* Horm. Behav., 38, 130–136.

Schank, J.C. and McClintock, M.K. (1992) *A coupled-oscillator model of ovarian-cycle synchrony among female rats*. J. Theor. Biol., 157, 317–362.

Schank, J.C. and McClintock, M.K. (1997) *Ovulatory pheromone shortens ovarian cycles of female rats living in olfactory isolation*. Physiol. Behav., 62, 899–904.

Schank, J.C., Tomasino, C.I. and McClintock, M.K. (1995) *The development of a pheromone isolation and delivery (PID) system for small mammals*. Anim. Tech., 46, 103–113.

Scheer, F.A. and Buijs, R.M. (1999) *Light affects morning salivary cortisol in humans*. J. Clin. Endocrinol. Metab., 84, 3395–3398.

Schellinck, H.M. and Brown, R.E. (1999) *Searching for the source of urinary odors of individuality in rodents*. In Johnston, R.E., Muller-Schwarze, D. and Sorensen, P.W. (eds.), Advances in chemical signals in vertebrates. Plenum, New York, pp. 267–280.

Schellinck, H.M., Forestel, C.A. and LoLordo, V.M. (2001) *A simple and reliable test of olfactory learning and memory in mice*. Physiol. Behav., 26, 663–672.

Schiffman, S.S., Sattely-Miller, E.A., Suggs, M.S. and Graham, B.G. (1995) *The effect of pleasant odors and hormone status on mood of women at midlife*. Brain Res. Bull., 36, 19–29.

Schleidt, M. (1980) *Personal odor and nonverbal communication*. Ethol. Sociobiol., 1, 225–231.

Schleidt, M., Hold, B. and Attili, G. (1981) *A cross-cultural study on the attitude towards personal odors*. J. Chem. Ecol., 7, 19–31.

Schultze-Westrum, T.G. (1969) *Social communication by chemical signals in flying phalangers* (Petaurus breviceps papuanus). In Pfaffman, C. (ed.), Olfaction and taste III. Rockefeller University Press, New York, pp. 268–277.

Scott, C.M., Marlin, D.J. and Schroter, R.C. (2001) *Quantification of the response of equine apocrine sweat glands to β2-adrenergic stimulation*. Equine Vet. J., 33, 605–612.

Semke, E., Distel, H. and Hudson, R. (1995) *Specific enhancement of olfactory receptor sensitivity associated with foetal learning of food odors in the rabbit*. Naturwissenschaften, 82, 148–149.

Shellinck, H.M., Slotnick, B.M. and Brown, R.E. (1997) *Odors of individuality originating from the major histocompatibility complex are masked by diet cues in the urine of rats*. Anim. Learn. Behav., 25, 193–199.

Shen, J., Niijima, A., Tanida, M., Horii, Y., Maeda, K. and Nagai, K. (2005) *Olfactory stimulation with scent of lavender oil affects autonomic nerves, lipolysis and appetite in rats*. Neurosci. Lett., 383, 188–193.

Shen, J., Niijima, A., Tanida, M., Horii, Y., Nakamura, T. and Nagai, K. (2007) *Mechanism of changes induced in plasma glycerol by scent stimulation with grapefruit and lavender essential oils*. Neurosci. Lett., 416, 241–246.

Sherborne, A.L., Thom, M.D., Paterson, S., Jury, F., Ollier, W.E.R., Stockley, P., Beynon, R.J. and Hurst, J.L. (2007) *The genetic basis of inbreeding avoidance in house mice.* Curr. Biol., 17, 2061–2066.

Shiferaw, B., Verrill, L., Booth, H., Zansky, S., Norton, D., Crim, S and Henao, O. (2008) *Are there gender differences in food consumption? The FoodNet Population Survey, 2006–2007.* Paper presented at the International Conference on Emerging Infectious Diseases. Atlanta, GA, March 19, 2008.

Shinohara, K., Morofushi, M., Funabashi, T. and Kimura, F. (2001) *Axillary pheromones modulate pulsatile LH secretion in humans.* Neuroreport, 12, 893–895.

Shinohara, K., Morofushi, M., Funabashi, T., Mitsushima, D. and Kimura, F. (2000) *Effects of 5α-androst-16-en-3α-ol on the pulsatile secretion of luteinizing hormone in human females.* Chem. Senses, 25, 465–467.

Shionoya, K., Moriceau, S., Bradstock, P. and Sullivan, R.M. (2007) *Maternal attenuation of hypothalamic paraventricular nucleus norepinephrine switches avoidance learning to preference learning in preweanling rat pups.* Horm. Behav., 52, 391–400.

Shorey, H.H. (1976) Animal communication by pheromones. Academic Press, New York.

Shors, T.J., Pickett, J., Wood, G. and Paczynski, M. (1999) *Acute stress persistently enhances estrogen levels in the female rat.* Stress, 3, 163–171.

Signoret, J.P. (1970) *Reproductive behaviour of pigs.* J. Reprod. Fertil., 11, 105–136.

Signoret, J.P. (1976) *Chemical communication and reproduction in domestic mammals.* In Doty, R.L. (ed.), Mammalian olfaction, reproductive processes, and behavior. Academic Press, New York, pp. 243–266.

Simons, R.R., Felgenhauer, B.E. and Jaeger, R.G. (1994) *Salamander scent marks—site of production and their role in territorial defense.* Anim. Behav., 48, 97–103.

Singer, A.G. (1991) *A chemistry of mammalian pheromones.* J. Steroid Biochem. Mol. Biol., 39, 627–632.

Singer, A.G., Agosta, W.C., Clancy, A.N. and Macrides, F. (1987) *The chemistry of vomeronasally detected pheromones: characterization of an aphrodisiac protein.* Ann. NY Acad. Sci., 519, 287–298.

Singer, A.G., Agosta, W.C., O'Connell, R.J., Pfaffmann, C., Bowen, D.V. and Field, F.H. (1976) *Dimethyl disulfide: an attractant pheromone in hamster vaginal secretion.* Science, 191, 948–950.

Singer, A.G., Clancy, A.N., Macrides, F. and Agosta, W.C. (1984) *Chemical studies of hamster vaginal discharge: effects of endocrine ablation and protein digestion on behaviorally active macromolecular fractions.* Physiol. Behav., 33, 639–643.

Singer, A.G., Clancy, A.N., Macrides, F., Agosta, W.C. and Bronson, F.H. (1988) *Chemical properties of a female mouse pheromone that stimulates gonado-tropin secretion in males.* Biol. Reprod., 38, 193–199.

Singer, A.G., Macrides, F., Clancy, A.N. and Agosta, W.C. (1986) *Purification and analysis of a proteinaceous aphrodisiac pheromone from hamster vaginal discharge.* J. Biol. Chem., 261, 13323–13326.

Singer, A.G., O'Connell, R.J., Macrides, F., Bencsath, A.F. and Agosta, W.C. (1983) *Methyl thiolbutyrate: a reliable correlate of estrus in the golden hamster.* Physiol. Behav., 30, 139–143.

Sipos, M.L., Kerchner, M. and Nyby, J.G. (1992) *An ephemeral sex pheromone in the urine of female house mice* (Mus domesticus). Behav. Neur. Biol., 58, 138–143.

Skandhan, K.P., Pandya, A.K., Skandhan, S. and Mehta, Y.B. (1979) *Synchronization of menstruation among inmates and kindreds.* Panminerva Med., 21, 131–134.

Slotnick, B.M. (1994) *The enigma of olfactory learning revisited.* Neurosci., 58, 1–12.

Slotnick, B.M. (2000) *Can rats acquire an olfactory learning set?* Behav. Neurosci., 114, 814–829.

Slotnick, B.M. (2001) *Animal cognition and the rat olfactory system.* Trends Cog. Sci., 5, 216–222.

Slotnick, B.M., Kufera, A. and Silberberg, A.M. (1991) *Olfactory learning and odor memory in the rat.* Physiol. Behav., 50, 555–561.

Smith, J.C. (1975) *Sound communication in rodents.* Symp. Zool. Soc. Lond., 37, 317–330.

Smith, T.D., Bhatnagar, K.P., Shimp, K.L., Kinzinger, J.H., Bonar, C.J., Burrows, A.M., Mooney, M.P., Siegel, M.I., Smith, T.D., Bhatnagar, K.P., Shimp, K.L., Kinzinger, J.H., Bonar, C.J., Burrows, A.M., Mooney, M.P. and Siegel, M.I. (2002) *Histological definition of the vomeronasal organ in humans and chimpanzees, with a comparison to other primates.* Anat. Rec., 267, 166–176.

Smotherman, W.P. (1982) *Odor aversion learning by the rat fetus.* Physiol. Behav., 29, 769–771.

Smotherman, W.P. and Robinson, S.R. (1987) *Prenatal expression of species-typical action patterns in the rat fetus* (Rattus norvegicus). J. Comp. Psychol., 101, 190–196.

Smotherman, W.P. and Robinson, S.R. (1990) *Rat fetuses respond to chemical stimuli in gas phase.* Physiol. Behav., 47, 863–868.

Snyder, R.L. and Taggart, N.E. (1967) *Effects of adrenalectomy on male induced pregnancy block in mice.* J. Reprod. Fertil., 14, 451.

Sobel, N., Prabhakaran, V., Hartley, C.A., Desmond, J.E., Glover, G.H., Sullivan,

E.V. and Gabrieli, J.D. (1999) *Blind smell: brain activation induced by an undetected air-borne chemical.* Brain, 122, 209–217.

Solem, J.H. (1966) *Plasma corticosteroids in mice.* Scand. J. Clin. Lab. Invest., 18, 1–36.

Sorensen, P.W. and Stacey, N.E. (1999) *Evolution and specialization of fish hormonal pheromones.* In Johnston, R.E., Müller-Schwarze, D. and Sorensen, P.W. (eds.), Advances in chemical signals in vertebrates. Kluwer Academic, New York, pp. 15–47.

Spencer, J., Gray, J. and Dalhouse, A. (1973) *Social isolation in the gerbil: its effect on exploratory or agonistic behavior and adrenocortical activity.* Physiol. Behav., 10, 231–237.

Spielman, A.I., Sunavala, G., Harmony, J.A.K., Stuart, W.D., Leyden, J.J., Turner, G., Vowels, B.R., Lam, W.C., Yang, S.J. and Preti, G. (1998) *Identification and immunohistochemical localization of protein precursors to human axillary odors in apocrine glands and secretions.* Arch. Derm., 134, 813–818.

Spironello, E. and deCatanzaro, D. (1999) *Sexual satiety diminishes the capacity of novel males to disrupt early pregnancy in inseminated female mice* (Mus musculus). J. Comp. Psychol., 113, 218–222.

Spironello-Vella, E. and deCatanzaro, D. (2001) *Novel male mice show gradual decline in the capacity to disrupt early pregnancy and in urinary excretion of testosterone and 17β-estradiol during the weeks immediately after castration.* Horm. Metab. Res., 33, 681–686.

Sprague, R.H. and Anisko, J.J. (1973) *Elimination patterns in the laboratory beagle.* Behaviour, 47, 257–267.

Stark, B. and Hazlett, B.A. (1972) *Effects of olfactory experience on aggression in* Mus musculus *and* Peromyscus maniculatus. Behav. Biol., 7, 265–269.

Stefanczyk-Krzymowska, S., Krzymowski, T., Grzegorzewski, W., Sowska, W. and Skipor, J. (2000) *Humoral pathway for local transfer of the priming pheromone androstenol from the nasal cavity to the brain and hypophysis in anaesthetized gilts.* Exp. Physiol., 85, 801–809.

Stehn, R.A. and Richmond, M.E. (1975) *Male-induced pregnancy termination in the prairie vole,* Microtus ochrogaster. Science, 1876, 1211–1213.

Stensaas, L.J., Lavker, R.M., Monti-Bloch, L., Grosser, B.I. and Berliner, D.L. (1991) *Ultrastructure of the human vomeronasal organ.* J. Steroid Biochem. Mol. Biol., 39, 553–560.

Stern, J.J. (1970) *Responses of male rats to sex odors.* Physiol. Behav., 5, 519–524.

Stern, K. and McClintock, M.K. (1998) *Regulation of ovulation by human pheromones.* Nature, 392, 177–179.

Stevenson, R.J. and Repacholi, B.M. (2003) *Age-related changes in children's hedonic response to male body odor.* Dev. Psychol., 39, 670–679.

Stickrod, G., Kimble, D.P. and Smotherman, W.P. (1982) *In utero taste/odor aversion conditioning in the rat.* Physiol. Behav., 28, 5–7.

Storey, A.E. (1986) *Influence of sires on male-induced pregnancy disruptions in meadow voles* (Microtus pennsylvanicus) *differs with stage of pregnancy.* J. Comp. Psychol., 100, 15–20.

Storey, A.E. and Snow, D.T. (1990) *Postimplantation pregnancy disruptions in meadow voles: relationship to variation in male sexual and aggressive behavior.* Physiol. Behav., 47, 19–25.

Stowers, L., Holy, T.E., Meister, M., Dulac, C. and Koentges, G. (2002) *Loss of sex discrimination and male-male aggression in mice deficient for TRP2.* Science, 295, 1493–1500.

Stowers, L. and Marton, T.F. (2005) *What is a pheromone? Mammalian pheromones reconsidered.* Neuron, 46, 699–702.

Strassmann, B.I. (1997) *The biology of menstruation in* Homo sapiens: *total lifetime menses, fecundity, and nonsynchrony in a natural-fertility population.* Curr. Anthro., 38, 123–129.

Strassmann, B.I. (1999) *Menstrual synchrony pheromones: cause for doubt.* Hum. Reprod., 14, 579–580.

Strott, C.A., Sundel, H. and Stahlman, M.T. (1975) *Maternal and fetal plasma progesterone, cortisol, testosterone and 17ß-estradiol in preparturient sheep: response to fetal ACTH infusion.* Endocrinology, 95, 1327–1322.

Stuart, A.M. (1970) *The role of chemicals in termite communication.* In Johnston, J.W., Jr., Moulton, D.G. and Turk, A. (eds.), Communication by chemical signals. Appleton-Century-Crofts, New York, pp. 79–106.

Suay, F., Salvador, A., Gonzalez-Bono, E., Sanchis, C., Martinez, M., Martinez-Sanchis, S., Simon, V.M. and Montoro, J.B. (1999) *Effects of competition and its outcome on serum testosterone, cortisol and prolactin.* Psychoneuro-endocrinology, 24, 551–566.

Swanson, H.H. and van de Poll, N.E. (1983) *Effects of an isolated or enriched environment after handling on sexual maturation and behaviour in male and female rats.* J. Reprod. Fertil., 69, 165–171.

Swingle, W.W., Fedor, E.J., Barlow, G., Collins, E.J. and Perlmutt, J. (1950) *Induction of pseudopregnancy in rat following adrenal removal.* Amer. J. Physiol., 167, 593–598.

Symonds, M.R. and Elgar, M.A. (2008) *The evolution of pheromone diversity.* Trends Ecol. Evol., 23, 220–228.

Takami, S., Getchell, M.L., Chen, Y., Monti-Bloch, L., Berliner, D.L., Stensaas, LJ and Getchell, T.V. (1993) *Vomeronasal epithelial cells of the adult human express neuron-specific molecules.* Neuroreport, 4, 375–378.

Tang, R., Webster, F.X. and Muller-Schwarze, D. (1995) *Neutral compounds from male castoreum of North American beaver*, Castor canadensis. J. Chem. Ecol., 21, 1745–1762.

Tanida, M., Niijima, A., Shen, J., Nakamura, T. and Nagai, K. (2005) *Olfactory stimulation with scent of essential oil of grapefruit affects autonomic neurotransmission and blood pressure.* Brain Res., 1058, 44–55.

Tanida, M., Niijima, A., Shen, J., Nakamura, T. and Nagai, K. (2006) *Olfactory stimulation with scent of lavender oil affects autonomic neurotransmission and blood pressure in rats.* Neurosci. Lett., 398, 155–160.

Tanida, M., Shen, J., Niijima, A., Yamatodani, A., Oishi, K., Ishida, N. and Nagai, K. (2008) *Effects of olfactory stimulations with scents of grapefruit and lavender oils on renal sympathetic nerve and blood pressure in Clock mutant mice.* Autonom. Neurosci., 139, 1–8.

Tanida, M., Yamatodani, A., Niijima, A., Shen, J., Todo, T. and Nagai, K. (2007) *Autonomic and cardiovascular responses to scent stimulation are altered in cry KO mice.* Neurosci. Lett., 413, 177–182.

Taylor, S.A. and Dewsbury, D.A. (1988) *Effects of experience and available cues on estrous versus diestrous preferences in male prairie voles*, Microtus ochrogaster. Physiol. Behav., 42, 379–388.

Taylor, S.A. and Dewsbury, D.A. (1990) *Male preferences for females of different reproductive conditions: a critical review.* In MacDonald, D.W., Müller-Schwarze, D. and Natynczuk, S.E. (eds.), Chemical signals in vertebrates 5. Oxford University Press, Oxford, pp. 184–198.

Teicher, M.H. and Blass, E.M. (1976) *Suckling in newborn rats: eliminated by nipple lavage, reinstated by pup saliva.* Science, 193, 422–425.

Teicher, M.H. and Blass, E.M. (1977) *First suckling response of the newborn albino rat: the roles of olfaction and amniotic fluid.* Science, 198, 635–636.

Terman, C.R. (1968) *Inhibition of reproductive maturation and function in laboratory populations of prairie deermice: a test of pheromone influence.* Ecology, 49, 1169–1172.

Terman, C.R. (1969) *Pregnancy failure in female prairie deermice related to parity and social environment.* Anim. Behav., 17, 104–108.

Terman, C.R. (1984) *Sexual development on female prairie deermice: influence of physical versus urine contact with grouped or isolated adult females.* J. Mammalogy, 65, 504–506.

Thiessen, D.D., Blum, S.L. and Lindzey, G. (1970) *A scent marking response associated with the ventral sebaceous gland of the Mongolian gerbil* (Meriones unguiculatus). Anim. Behav., 18, 26–30.

Thiessen, D.D., Friend, H.C. and Lindzey, G. (1968) *Androgen control of territorial marking in the Mongolian gerbil.* Science, 160, 432–434.

Thiessen, D.D., Regnier, F.E., Rice, M., Goodwin, M., Isaacks, N. and Lawson, N.

(1974) *Identification of a ventral scent marking pheromone in the male Mongolian gerbil* (Meriones unguiculatus). Science, 184, 83–85.

Thom, M.D., Stockley, P., Jury, F., Ollier, W.E.R., Beynon, R.J. and Hurst, J.L. (2008) *The direct assessment of genetic heterozygosity through scent in the mouse.* Curr. Biol., 18, 619–623.

Thomas, D.A., Riccio, D.C. and Myer, J.S. (1977) *Age of stress-produced odorants and the Kamin effect.* Behav. Biol., 20, 433–440.

Thompson, W.F., Schellenberg, E.G. and Husain, G. (2001) *Arousal, mood, and the Mozart effect.* Psychol. Sci., 12, 248–251.

Thorne, F., Neave, N., Scholey, A., Moss, M. and Fink, B. (2002) *Effects of putative male pheromones on female ratings of male attractiveness: influence of oral contraceptives and the menstrual cycle.* Neuroendocr. Lett., 23, 291–297.

Thysen, B., Elliott, W.H. and Katzman, P.A. (1968) *Identification of estra-1,3,5(10),16-tetraen-3-ol (estratetraenol) from the urine of pregnant women (1).* Steroids, 11, 73–87.

Tirindelli, R., Mucignat-Caretta, C. and Ryba, N.J. (1998) *Molecular aspects of pheromonal communication via the vomeronasal organ of mammals.* Trend Neurosci., 21, 482–486.

Tisserand, R. (1993) The art of aromatherapy. C.W. Daniel, Essex, UK.

Titchener, E.B. (1928) A text-book of psychology. Macmillan, New York.

Tobin, D.J. (2006) *Biochemistry of human skin—our brain on the outside.* Chem. Soc. Rev., 35, 52–67.

Todrank, J., Heth, G. and Johnston, R.E. (1999) *Social interaction is necessary for discrimination between and memory for odours of close relatives in golden hamsters.* Ethology, 105, 771–782.

Torner, L., Toschi, N., Pohlinger, A., Landgraf, R. and Neumann, I.D. (2001) *Anxiolytic and anti-stress effects of brain prolactin: improved efficacy of antisense targeting of the prolactin receptor by molecular modeling.* J. Neurosci., 21, 3207–3214.

Treloar, A.E., Boynton, R.E., Behn, B.G. and Brown, B.W. (1967) *Variation of the human menstrual cycle through reproductive life.* Internat. J. Fertil., 13, 77–126.

Trevathan, W.R., Burleson, M.H. and Gregory, W.L. (1993) *No evidence for menstrual synchrony in lesbian couples.* Psychoneuroendocrinology, 18, 425–435.

Trotier, D., Eloit, C., Wassef, M., Talmain, G., Bensimon, J.L., Doving, K.B. and Ferrand, J. (2000) *The vomeronasal cavity in adult humans.* Chem. Senses, 25, 369–380.

Valenta, J.G. and Rigby, M.K. (1968) *Discrimination of the odor of stressed rats.* Science, 161, 599–601.

Vandenbergh, J.G. (1967) *Effect of the presence of a male on the sexual maturation of female mice.* Endocrinology, 81, 345–349.

Vandenbergh, J.G. (1969) *Male odor accelerates female sexual maturation in mice.* Endocrinology, 84, 658–660.

Vandenbergh, J.G., Drickamer, L.C. and Colby, D.R. (1972) *Social and dietary factors in the sexual maturation of female mice.* J. Reprod. Fertil., 28, 397–405.

Vandenbergh, J.G., Whitsett, J.M., and Lombardi, J.R. (1975) *Partial isolation of a pheromone accelerating puberty in female mice.* J. Reprod. Fertil., 43, 515–523.

Van der Lee, S. and Boot, L.M. (1955) *Spontaneous pseudopregnancy in mice.* Acta Physiol. Pharmacol. Neerlandica, 4, 442–443.

Van der Lee, S. and Boot, L.M. (1956) *Spontaneous pseudopregnancy in mice. II.* Acta Physiol. Pharmacol. Neerlandica, 5, 213–214.

Vasilieva, N.Y., Cherepanova, E.V., von Holst, D. and Apfelbach, R. (2000) *Predator odour and its impact on male fertility and reproduction in* Phodopus campbelli *hamsters.* Naturwissenschaften, 87, 312–314.

Wallace, P., Owen, K. and Thiessen, D.D. (1973) *The control and function of maternal scent marking the Mongolian gerbil.* Physiol. Behav., 10, 463–466.

Wallis, J. (1985) *Synchrony of estrous swelling in captive group-living chimpanzees (Pan-Troglodytes).* Int. J. Primatol., 6, 335–350.

Wang, H.W., Wysocki, C.J. and Gold, G.H. (1993) *Induction of olfactory receptor sensitivity in mice.* Science, 260, 998–1000.

Wang, J., Eslinger, P.J., Smith, M.B. and Yang, Q.X. (2005) *Functional magnetic resonance imaging study of human olfaction and normal aging.* J. Gerontol. A Biol. Sci. Med. Sci., 60, 510–514.

Warrenburg, S. (2002) *Measurement of emotion in olfactory research.* Chem. Taste Mech. Behav. Mimics, 825, 243–260.

Watson, D., Clark, L.A. and Tellegen, A. (1988) *Development and validation of brief measures of positive and negative affect: the PANAS scales.* J. Pers. Soc. Psychol., 54, 1063–1070.

Watson, M., Clulow, F.V. and Mariotti, F. (1983) *Influence of olfactory stimuli on pregnancy of the meadow vole,* Microtus pennsylvanicus, *in the laboratory.* J. Mammal, 64, 706–708.

Watson, R.H. and Radford, H.M. (1960) *The influence of rams on onset of oestrus in Merino ewes in the spring.* Aust. J. Agric. Res., 11, 65–71.

Weir, M.W. and DeFries, J.C. (1963) *Blocking of pregnancy in mice as a function of stress.* Psychol. Rep., 13, 365–366.

Weizenbaum, F., McClintock, M. and Adler, N. (1977) *Decreases in vaginal acyclicity of rats when housed with female hamsters.* Horm. Behav., 8, 342–347.

Weller, A. and Weller, L. (1992) *Menstrual synchrony in female couples.* Psychoneuroendocrinology, 17, 171–177.

Weller, A. and Weller, L. (1993a) *Menstrual synchrony between mothers and daughters and between roommates.* Physiol. Behav., 53, 943–949.

Weller, L. and Weller, A. (1993b) *Multiple influences of menstrual synchrony: kibbutz roommates, their best friends, and their mothers.* Amer. J. Human Biol., 5, 173–179.

Weller, A. and Weller, L. (1995a) *Examination of menstrual synchrony among women basketball players.* Psychoneuroendocrinology, 20, 613–622.

Weller, A. and Weller, L. (1995b) *The impact of social interaction factors on menstrual synchrony in the workplace.* Psychoneuroendocrinology, 20, 21–31.

Weller, A. and Weller, L. (1997a) *Menstrual synchrony under optimal conditions: Bedouin families.* J. Comp. Psychol., 111, 143–151.

Weller, L. and Weller, A. (1997b) *Menstrual variability and the measurement of menstrual synchrony.* Psychoneuroendocrinology, 22, 115–128.

Weller, A. and Weller, L. (1998) *Prolonged and very intensive contact may not be conductive to menstrual synchrony.* Psychoneuroendocrinology, 23, 19–32.

Weller, L., Weller, A. and Avinir, O. (1995) *Menstrual synchrony: only in roommates who are close friends?* Physiology & Behavior, 58, 883–889.

Weller, L., Weller, A., Koresh-Kamin, H. and Ben Shoshan, R. (1999a) *Menstrual synchrony in a sample of working women.* Psychoneuroendocrinology, 24, 449–459.

Weller, L., Weller, A. and Roizman, S. (1999b) *Human menstrual synchrony in families and among close friends: examining the importance of mutual exposure.* J. Comp. Psychol., 113, 261–268.

Wells, D.L. and Hepper, P.G. (2003) *Directional tracking in the domestic dog, Canis familiaris.* Appl. Anim. Behav. Sci., 84, 297–305.

Wever, R.A. (1979) The circadian system of man" results of experiments under temporal isolation. Springer-Verlag, New York.

White, N.R., Prasad, M., Barfield, R.J. and Nyby, J.G. (1998) *40- and 70-kHz vocalizations of mice (Mus musculus) during copulation.* Physiol. Behav., 63, 467–473.

Whitney, G. (1973) *Vocalization of mice influenced by a single gene in a heterogeneous population.* Behav. Genet., 3, 57–64.

Whitney, G., Alpern, M., Dizinno, G. and Horowitz, G. (1974) *Female odors evoke ultrasounds from male mice.* Anim. Learn. Behav., 2, 13–18.

Whitney, G., Coble, J.R., Stockton, M.D. and Tilson, E.F. (1973) *Ultrasonic emissions: do they facilitate courtship of mice.* J. Comp. Physiol. Psychol., 84, 445–452.

Whitten, W. (1999) *Reproductive biology: pheromones and regulation of ovulation.* Nature, 401, 232–233.

Whitten, W.K. (1959) *Occurrence of anoestrus in mice caged in groups.* J. Endocrinol., 18, 102–107.

Whitten, W.K. (1966) *Pheromones and mammalian reproduction.* In McLaren,

A. (ed.), Advances in reproductive physiology. Academic Press, New York, pp. 159–177.

Whitten, W.K. (1975) *Responses to pheromones by mammals*. In Denton, D.A. and Coghlan, J.P. (eds.), Olfaction and taste V. Academic Press, New York, pp. 389–395.

Whitten, W.K. and Bronson, F.H. (1970) *The role of pheromones in mammalian reproduction*. In Johnston, J.W., Jr., Moulton, D.G. and Turk, A. (eds.), Communication by chemical signals. Appleton-Century-Crofts, New York, pp. 309–325.

Whitten, W.K. and Champlin, A.K. (1972) *Bibliography (with review) on phero-mones & olfaction in mammalian reproduction*. Biblio. Reprod., 19, 149–156, 297–303.

Whitten, W.K. and Champlin, A.K. (1973) *The role of olfaction in mammalian reproduction*. Handbook of physiology. Section 7: Endocrinology. American Physiological Society, Washington, DC, pp. 109–123.

Wierson, M., Long, P.J. and Forehand, R.L. (1993) *Toward a new understanding of early menarche: the role of environmental stress in pubertal timing*. Adolescence, 28, 913–924.

Wilke, K., Martin, A., Terstegen, L. and Biel, S.S. (2007) *A short history of sweat gland biology*. Int. J. Cosmet. Sci., 29, 169–179.

Williams, G.W., McGinnis, M.Y. and Lumia, A.R. (1992) *The effects of olfactory bulbectomy and chronic psychosocial stress on serum glucocorticoids and sexual behavior in female rats*. Physiol. Behav., 52, 755–760.

Wilson, D.A. and Stevenson, R.J. (2006) Learning to smell: Olfactory perception from neurobiology to behavior. Johns Hopkins University Press, Baltimore.

Wilson, E.O. (1963) *Pheromones*. Sci. Amer., 208, 100–114.

Wilson, E.O. (1970) *Chemical communication within animal species*. In Sondheimer, E. and Simeone, J.B. (eds.), Chemical ecology. Academic Press, New York, pp. 133–155.

Wilson, E.O. (1972) *Animal communication*. Sci. Amer., 227, 53–61.

Wilson, E.O. and Bossert, W.H. (1963) *Chemical communication among animals*. Rec. Prog. Horm. Res., 19, 673–710.

Wilson, H.C. (1987) *Female axillary secretions influence women's menstrual cycles: a critique*. Horm. Behav., 21, 536–546.

Wilson, H.C. (1992) *A critical review of menstrual synchrony research*. Psychoneuroendocrinology, 17, 565–591.

Wilson, H.C. (1993) *Reply to letter by Graham*. Psychoneuroendocrinology, 18, 535–539.

Wilson, H.C., Kiefhaber, S.H. and Gravel, V. (1991) *Two studies of menstrual synchrony: negative results*. Psychoneuroendocrinology, 16, 353–359.

Winans, S.S. and Powers, J.B. (1974) *Neonatal and two-stage olfactory bulbectomy: effects on male hamster sexual behavior.* Behav. Biol., 10, 461–471.

Winans, S.S. and Powers, J.B. (1977) *Olfactory and vomeronasal deafferentation of male hamsters: histological and behavioral analyses.* Brain Res., 126, 325–344.

Winman, A. (2004) *Do perfume additives termed human pheromones warrant being termed pheromones?* Physiol. Behav., 82, 697–701.

Wirsig, C.R. and Leonard, C.M. (1987) *Terminal nerve damage impairs the mating behavior of the male hamster.* Brain Res., 417, 292–303.

Witt, M., Georgiewa, B., Knecht, M. and Hummel, T. (2002) *On the chemosensory nature of the vomeronasal epithelium in adult humans.* Histochem. Cell Biol., 117, 493–509.

Wyatt, T.D. (2008) Pheromones and animal behavior: communication by smell and taste. Cambridge University Press, Cambridge.

Wynne-Edwards, V.C. (1962) Animal dispersion in relation to social behaviour. Oliver and Boyd, Ltd., Edinburgh.

Wysocki, C.J. (1979) *Neurobehavioral evidence for the involvement of the vomeronasal system in mammalian reproduction.* Neurosci. Biobehav. Rev., 3, 301–341.

Yamazaki, K., Beauchamp, G.K., Kupniewski, D., Bard, J., Thomas, L. and Boyse, E.A. (1988) *Familial imprinting determines H-2 selective mating preferences.* Science, 240, 1331–1332.

Yamazaki, K., Boyse, E.A., Mike, V., Thaler, H.T., Mathieson, B.J., Abbott, J., Boyse, Zayas, Z.A. and Thomas, L. (1976) *Control of mating preferences in mice by genes in the major histocompatibility complex.* J. Exp. Med., 144, 1324–1335.

Yamazaki, K., Singer, A. and Beauchamp, G.K. (1998) *Origin, functions and chemistry of H-2 regulated odorants.* Genetica, 104, 235–240.

Yang, Z.W. and Schank, J.C. (2006) *Women do not synchronize their menstrual cycles.* Human Nature, 17, 433–447.

Yerkes, R.M. (1911) Introduction to psychology. H. Holt, New York.

Yoneda, Y., Kanmori, K., Ida, S. and Kuriyama, K. (1983) *Stress-induced alteration in metabolism of aminobutyric acid in rat brain.* J. Neurochem., 40, 350–356.

Yoshimura, H. (1980) *Cholinergic mechanisms in scent marking behavior by Mongolian gerbils* (Meriones unguiculatus). Pharmacol. Biochem. Behav., 13, 519–523.

Youngentob, S.L. and Kent, P.F. (1995) *Enhancement of odorant-induced mucosal activity patterns in rats trained on an odorant identification task.* Brain Res., 670, 82–88.

Zacharias, R., de, C.D. and Muir, C. (2000) *Novel male mice disrupt pregnancy despite removal of vesicular-coagulating and preputial glands.* Physiol. Behav., 68, 285–290.

Zalaquett, C. and Thiessen, D. (1991) *The effects of odors from stressed mice on conspecific behavior.* Physiol. Behav., 50, 221–227.

Zeng, X.N., Leyden, J.J., Brand, J.G., Speilman, A.I., McGinley, K.J. and Preti, G. (1992) *An investigation of human apocrine gland secretion for axillary odor precursors.* J. Chem. Ecol., 18, 1039–1055.

Zeng, X.N., Leyden, J.J., Lawley, H.J., Sawano, K., Nohara, I. and Preti, G. (1991) *Analysis of characteristic odors from human male axillae.* J. Chem. Ecol., 17, 1469–1491.

Zeng, X.N., Leyden, J.J., Spielman, A.I. and Preti, G. (1996) *Analysis of characteristic human female axillary odors: qualitative comparison to males.* J. Chem. Ecol., 22, 237–257.

Zinner, D., Schwibbe, M.H. and Kaumanns, W. (1994) *Cycle synchrony and probability of conception in female Hamadryas baboons Papio-Hamadryas.* Behav. Ecol. Sociobiol., 35, 175–183.

Ziomkiewicz, A. (2006) *Menstrual synchrony: fact or artifact?* Hum. Nature, 17, 419–432.

Zondek, B. and Tamari, I. (1967) *Effects of auditory stimuli on reproduction.* Ciba Foundation Study Group, 26, 4–19.

Zufall, F. and Leinders-Zufall, T. (2007) *Mammalian pheromone sensing.* Curr. Opinion Neurobiol., 17, 483–489.

King, J.A., 76, 106
King, M.G., 94
Kippin, T.E., 43
Kirk-Smith, M.D., 141–143, 145, 149
Kloek, J., 139
Knecht, M., 129
Knight, T.W., 99
Knutson, B., 70
Kodis, M., 186
Koelega, H.S., 140
Kondo, Y., 91, 92
Kouros-Mehr, H., 130
Krishnamurthy, V., 48–49
Kroner, C., 66
Krout, K.E., 164
Kruse, S.M., 67–68
Krutova, V.I., 41
Krzymowski, T., 61
Kurt, F., 94
Kwan, T.K., 139

Labov, J.B., 124
Labows, J.N., 140
Laing, D.G., 50, 94–95, 133
Lalumiere, M.L., 146
Lamond, D.R., 115
Lang, P.J., 166
Larson, J., 41
Lehrman, D., 9
Leidahl, L.C., 78, 79
Le Magnen, J., 139
Lent, P., 94
Leon, M., 78, 79
Leshem, M., 106
Levai, O., 27
Leyden, J.J., 126
Leypold, B.G., 29, 75
Liman, E.R., 130
Little, B.B., 174
Lloyd-Thomas, A., 103
Lorenz, K., 9, 14, 189–190
Lott, D.F., 203n4
Ludvigson, H.W., 95, 107
Lundström, J.N., 30, 155–156
Luo, M., 28, 77
Lüscher, M., 9, 13, 16
Lydell, K., 55, 68, 69

Ma, W., 115, 116, 119
Mackay-Sim, A., 94–95
Mackintosh, J.H., 78
MacNiven, E., 105, 106
Macrides, F., 39, 63
Maggio, J.C., 72
Maier, I., 16
Mainardi, D., 39, 40
Mar, A., 107
Marchlewska-Koj, A., 105, 116, 117, 118–119, 120
Marr, J.N., 40
Marsden, H.M., 105
Marshall, D.A., 36
Martinez-Ricos, J., 19
Marton, T.F., 29
Maruniak, J.A., 45, 109, 112, 113
Mateo, J.M., 41
Matsumoto-Oda, A., 123
Matteo, S., 174
Mayr, E., 190–191
Mazur, A., 34
McCarty, R., 39
McCaul, K.D., 34
McClintock, M.K., 18, 120, 121–122, 153–155, 168–170, 171, 177–182
McClure, T.J., 105
McCoy, N.L., 159, 160–161
McDougall, W., 9
McNair, D.M., 153, 166
McNeilly, A.S., 120
Meisami, E., 130, 205n2
Melrose, D.R., 12, 61
Mennella, J.A., 37, 39
Meredith, M., 17, 27, 28, 84, 129, 131
Michael, R.P., 12, 57, 58, 147
Milenkovic, L., 105, 116
Mody, J.K., 117
Mogg, K., 166
Moncho-Bogani, J., 35, 42, 92, 93
Monder, H., 108
Monti-Bloch, L., 125, 130, 131, 153, 154
Montigny, D., 91
Moran, D.T., 28, 129, 131
Moriceau, S., 39
Morris, N.M., 57

Winans, S.S., 62
Winberg, J., 38
Winman, A., 161, 162
Wirsig, C.R., 62
Witt, M., 130
Wolff, L.G., 119
Wyatt, T.D., 15–16
Wynne-Edwards, V.C., 7
Wysocki, C.J., 27

Yerkes, R.M., 9
Yoneda, Y., 35

Yoshimura, H., 203n8
Youngentob, S.L., 38

Zacharczuk-Kakietek, M., 119
Zacharias, R., 103
Zahorik, D.M., 63
Zalaquett, C., 95
Zeng, X.N., 134, 140
Zinner, D., 123
Zondek, B., 106
Zufall, F., 29

Subject Index

adrenocorticotropic hormone (ACTH), 105, 119
aliphatic acids, 57. *See also* copulin
alloiohormones, 9
androstadienone (AND), 130–132, 140–141; as influence on psychological and physiological responses, 153–156
androstenedione, 103
androstenol, 59, 60, 61; as influence on evaluation of others, 146–151; as influence on mood, 153, 154–155; as influence on social behavior, 151–152; as pheromone, 150
androstenone: as influence on evaluation of others, 146–151; as influence on seating choices, 141–46; as pheromone, 59–60, 61, 138–151
androsterone, 103
antelope. *See* pronghorn antelope
antidepressants: and the Bruce effect, 107
aphrodisin, 65–66
apocrine glands, 126–127, 132, 134. *See also* axillary secretions
Asian elephants. *See* elephants
Athena Institute, 160, 161
Athena Pheromone 10:13, 160
Athena Pheromone 10X, 160, 162
auditory stimulation: effect of on animals, 163; effect of on humans, 168
axillary secretions: effect of diet on, 135; mood as influenced by, 137–138; odors of as sex attractant, 132–137

beavers: scent marking by, 49
boars: lordosis facilitation in, 60–62; and "releasing" pheromones, 59–62
bombykol, 10, 23–24
brain imaging: as applied to human pheromone studies, 157–159
breastfeeding: pheromones associated with, 183–184
Bruce effect, 36, 102–109, 119; blocking of, 104, 107; effect of hormones on, 107–109; physiological basis for, 103–105; role of adrenal gland in, 105; stress as factor in, 105–108, 202n1

cecotroph, 78–79
chemical signal, 24
chemosensory learning, 35–46. *See also* olfactory learning
chimpanzees: estrous synchrony in, 123–124
civetone, 11, 17–18
cognitive performance: effect of odors on, 165
cognitive processes: as mediated by chemically induced responses, 34–35; and scent marking, 55
color: physiological and psychological effect of, 168
copulin, 57–59; as influence on evaluation of others, 146–151; as influence on social behavior, 151–152
corticosterone, 116, 117, 119

dehydroepiandrosterone, 103
dexamethasone, 117
diet: effect of on axillary secretions, 135; effect of on putative pheromones, 97; as factor in scent marks, 49–50
dimethyl disulfide (DMDS), 63–65
dodecyl propionate, 46
dogs: and discernment of scent, 54–55; and "releasing" pheromones, 66–68

eccrine sweat glands, 127–128
ectohormone, 9
elephants: frontalin as pheromone in males, 82–86; and "releasing" pheromones, 81–82
epiadrosterone, 103
epinephrine, 104
estratetraenol (EST), 131; as influence on psychological and physiological responses, 153–156
estrogen, 104–105, 202n2
estrous synchrony: in nonhuman primates, 123–124; in rats, 120–123
estrus induction: in mice, 118–120
ether stress, 119
eugenol, 168

female hamster vaginal secretions (FHVS), 62–63, 201–202n1; and aphrodisin, 65–66; and dimethyl disulfide, 63–65
flank gland secretions: of hamsters, 43
flehmen response: in elephants, 81–82, 83–86
frontalin: elephants' responses to, 82–86
functional magnetic resonance imaging (fMRI): and human pheromone studies, 159

gamone, 197n1
gerbils: olfactory learning in, 39; and "releasing" pheromones, 73–75
glands. See apocrine glands; eccrine sweat glands; sebaceous glands
golden lion tamarin: estrous synchrony in, 123
grapefruit oil, 163, 164
Grüneberg ganglion, 95–96
guinea pigs: scent marking by, 49

haloperidol, 116
Hamadryas baboon: estrous synchrony in, 123
hamsters: and FHVS, 62–66, 201–202n1; olfactory learning in, 42–43; and "releasing" pheromones, 62–66; scent marking by, 54, 55
homoiohormones, 9
hormone levels: changes in response to odors, 43–44; changes resulting from cognitive processes, 34–35
hormone replacement therapy, 161, 167
hormones: behavioral responses to, 21–22; and the Bruce effect, 107–109; pheromones as, 7–8, 13–15. See also estrogen; luteinizing hormone; testosterone; and names of other specific hormones
human pheromones (putative): androstenones and related agents as, 138–151; axillary secretions as source of, 132–138; brain imaging applied to studies of, 157–159; in breastfeeding women, 183–184; evidence for, 125–26; as political issue, 29–31; potential receptors for, 128–132; problems with concept of, 125–126, 141, 152, 154, 159, 168, 182–183, 184; role of in menstrual synchrony, 173–183; sources of, 126–128; synthesized, 159–162; and the vomeronasal organ, 128–132
humans: effect of odors on mood and behavior, 164–168; menstrual synchrony in, 123; and olfactory learning, 37, 38; pleasant odors as stress reducer in, 107; stress as factor in onset of puberty in, 114; vomeronasal organ in, 28, 125. See also breastfeeding; menstrual synchrony

imprinting pheromones, 22–23
insects: and chemicals employed in communication, 8–10, 191–192; pheromones in, 8–9, 22, 197n2
instinct, concept of, 9, 189–190
isoamylamine, 110
isobutylamine, 110
isovaleric acid, 58, 87–88

Jacobson's organ. *See* vomeronasal organ (VNO)

lavender oil, 163
L-Dopa, 119
learning. *See* olfactory learning
Lee-Boot effect: explanation for, 115–116; generalizability of to other species, 117; in mice, 114–118
lemon oil, 165–166
lordosis behavior: in boars, 60–62
luteinizing hormone (LH), 34; in ewes, 99–100; release of, 101; and sexual experience, 99–101

major histocompatibility complex (MHC), 198n6, 199–200n4
major urinary protein (MUP), 111–112, 198n6; role of in aggressive behavior, 76, 77–78
mammal behavior: complexity of, 3–4; experience as factor in, 22–23, 40; and olfactory learning, 32–46; scent marking, 47–55; stress-related odors in, 94–96. *See also* olfactory learning; pheromones, mammalian; *and names of specific mammals*
mammalian secretions and excretions: chemical complexity of, 23–25, 96–97
maternal pheromone: in rats, 78–81
meadow voles: scent marking by, 49
medial amygdala: in rats, 92
menstrual synchrony, 168–173; axillary influences on, 174–178; criteria for, 169–170; methodological problems inherent in studies of, 170–173, 174, 179–182; putative role of pheromones in, 173–183
metatarsal glands, 50
methyl *p*-hydroxybenzoate, 66–68
methyl thiobutyrate (MTB), 64
mice: aggressive behavior in, 75–78; and the Bruce effect, 102–109; as compared with insects, 197–198n4; and estrus induction, 118–120; Lee-Boot effect in, 114–118; olfactory learning in, 38, 39–40, 42; pregnancy-blocking phenomenon in, 102–109; and "priming" pheromones, 101; pubertal accel-

eration in, 109–114; and "releasing" pheromones, 69–73, 75–78, 92–94; scent marking by, 44–46; ultrasound vocalizations of, 69–73; urinary proteins in, 75–78
Mongolian gerbils. *See* gerbils
mood: androstenol as influence on, 153, 154–155; axillary secretions as influence on, 137–138; effect of odors on, 165–167
mosaic signal, 24
murine nervous system, 197–198n4, 198n6. *See also* mice; rats
muscone: as influence on mood, 154–155
music. *See* auditory stimulation
musketone, 11
musks, 11–12
musth: in elephants, 82–83

neonatal olfactory learning, 38–41; in gerbils, 39; in humans, 38–39; in rats and mice, 38, 39–40
nervus terminalis (CN O), 62
neuro-specific enolase (NSE), 131
nipple search response, 88–91
nonhuman primates: estrous synchrony in, 123–124

odorants: as distinguished from pheromones, 35–36; effect of on neural activity in animals, 163–164; effect of on neural activity in humans, 164–168; impact of on social behavior, 32; and piloerection, 94
odor mosaic, 24
odors: experience as factor in responses to, 40, 41–42, 200n6; meanings of, 43–46; and nest development of birds, 200n5; and nesting preferences of mice, 201n8; as stress alleviator, 107; as response to stress, 94–96; role of in species recognition, 200–201n7. *See also* olfactory learning; olfactory system
olfactory learning, 22–23, 190–191, 199nn2–3; in adults, 41–46; in elephants, 85; evidence for, 36–46; in humans, 37, 38, 41; and mammal

behavior, 32–46; neonatal, 38–41, 78–81, 89, 191; in rabbits, 89; in the uterus, 37–38, 191
olfactory system: complexity of, 33, 188; criteria for, 205n2; evolution of, 41; in mammals, 28–29; and pheromones, 13, 14–15. *See also* odorants; olfactory learning; vomeronasal organ (VNO)
opioids, 199n1

phenylacetic acid, 73–75
pheromone blend, 24
pheromones: in algae, 16; alternative terminology for, 197n2, 198n4; behavioral, 17–18; criteria for, 20–21, 84, 89; defined as olfactory stimulus, 13, 14–15, 16–17, 46; definitions of, 12–19, 185–186, 188–189; dictionary definitions of, 13–14; evolutionary argument for, 6–8; functions of, 14–15; as hormones, 13–15; in insects, 8–9, 22, 197n2; origin of concept of, 8–10; popular definitions of, 186; scientists' definitions of, 12–13, 16–19; as species-specific, 14–15, 16–19; textbook definitions of, 14–16; usefulness of term, 3–4, 9–10, 46, 186, 189, 197n2; volatility of, 13–14, 25–26. *See also* human pheromones (putative); pheromones, mammalian
pheromones, mammalian: alternative approaches to conceptualizing of, 193–194; and chemical complexity, 12, 23–25, 30–31, 193; chemical evidence of, 25–26; chemical isolation of, 12; concept of, 5–12; difficulty of defining, 24–25; as distinguished from odorants, 35–36; early concerns regarding, 19–23; and interspecific interaction, 194; learning as consideration in assessment of, 22–23; maternal, 78–81; as non-odors, 25–26; as political issue, 29–31; problems with concept of, 1–4, 32–33, 41, 46, 53–54, 56, 57–58, 62, 66, 68, 69, 72–73, 75, 76–77, 79–81, 82, 85–86, 88, 92, 93–94, 96–97, 98–99, 108–109, 114, 118, 124, 185–94; urine as

putative source of, 59, 71–73, 76–78, 81, 91, 92–93, 98, 101, 103–104, 108, 109–112, 119, 124, 126; vaginal secretions as putative source of, 57–59, 62–66, 204–205n1; volatility of, 25–26; and the vomeronasal organ, 26–29, 30, 61. *See also* human pheromones (putative); pheromones; "priming" pheromones; "releasing" pheromones
phylogenetic scale, 33
piloerection, 94
poisoned partner effect, 44
positron emission tomography (PET): and human pheromone studies, 157–159
pregnancy blocking: in mice, 102–109
primates. *See* humans; nonhuman primates
"priming" pheromones, 2, 19, 98–99, 189; and the Bruce effect, 102–109; and estrous synchrony, 120–123; and estrus induction, 118–120; and the Lee-Boot effect, 114–118; and mice, 101, 109–120; and nonhuman primates, 123–124; and pubertal acceleration, 109–114; and rams, 99–101; and rats, 120–123
progesterone, 103, 119
prolactin, 116–117
pronghorn antelope: and "releasing" pheromones, 86–88; subauricular glands of, 86–88
pubertal acceleration: in mice, 109–114

rabbits: nipple search behavior in, 88–91; olfactory learning in, 89; and "releasing" pheromones, 88–91
Raoult's law, 25
rats: as compared with insects, 197–198n4; effect of odorants on, 163–164; estrous females and erection-elicitation, 91–92; estrous synchrony in, 120–123; food preferences of, 44; maternal pheromone proposed for, 78–81; olfactory conditioning in, 43–44; olfactory learning in, 38, 39–40; and "releasing" pheromones, 68–69, 91–92; stress response in, 94–96

"releasing" pheromones, 2, 17, 96–97, 189, 197–198n4; and aggressive behavior in mice, 75–78; and boars, 59–62; concerns about concept of, 19–20, 56; and dogs, 66–68; and elephants, 81–86; and gerbils, 73–75; and hamsters, 62–66; and learning, 36; and mice, 69–73, 75–78, 92–94; and rats, 68–69, 91–92; and rhesus monkeys, 57–59

reserpine, 106–107

rhesus monkeys: and "releasing" pheromones, 57–59

Rodentia: and olfactory learning, 22–23, 37–38; pheromones in, 5–6. *See also* gerbils; hamsters; mice; rabbits; rats

scala naturae, 33

scent glands: components of, 48, 49, 50–51

scent marking, 47–55, 201nn1–2; by beavers, 49; complexity of, 50, 54–55, 190; and dietary factors, 49–50; functions of, 50–55; by hamsters, 54, 55; by meadow voles, 49; by mice, 44–46

sebaceous glands, 128

sheep: and "priming" pheromones, 99–101

silkworm moths, 10, 11

steroids. *See* androstenedione; androstenol; androstenone; androsterone; estratetraenol; estrogen; muscone; progesterone; testosterone

stress, 73; effect of pleasant odors on, 107; as factor in the Bruce effect, 105–108, 202n1; as factor in the Lee-Boot effect, 115–117; odors associated with, 94–96; and onset of puberty, 113–114, 193; and social isolation, 203n8

synthesized human sex pheromones: chemical composition of, 160; flaws in studies of, 160–161, 162; as perfume or cologne, 159–162; studies of, 159–162

tarsal glands, 50

territorial marking. *See* scent marking

testosterone: and the Bruce effect, 103, 107–109

testosterone levels, 34–35

2-methylbut-2-enal, 89–91

2-(*sec*-butyl)thiazoline, 76–78

2,3-dehydro-*exo*-brevicomin, 76–78

ultrasound vocalizations: in mice, 69–73, 202n4

underarms. *See* axillary secretions

urine: components of, 48–49, 101; learned responses to odor of, 23; and "priming" pheromones, 98; and pubertal acceleration, 109–112; as putative source of pheromones, 59, 71–73, 76–78, 81, 91, 92–93, 98, 101, 103–104, 108, 109–112, 119, 124, 126; and scent marking, 47–49. *See also* major urinary protein (MUP) complex

uterus: olfactory learning in, 37–38, 191

vaginal pulse amplitude (VPA): effect of fragrances on, 167

vaginal secretions: of hamsters, 43, 62–66, 201–202n1; of humans, 204–205n1; as putative source of pheromones, 57–59, 62–66, 204–205n1; of rhesus monkeys, 57–59

Vandenbergh effect, 109–114

vocalizations. *See* ultrasound vocalizations

vomeronasal organ (VNO), 18, 23, 58, 198nn5–6; and the Bruce effect, 103–104; components of, 205n2; effect of damage to or removal of, 28, 62, 75, 92, 101, 115; and female mouse urine, 101; in humans, 28, 125, 129–130; and the Lee-Boot effect, 115; location of, 26–27; and male mouse urine, 92–93; as mediator of pheromonal activity in humans, 128–32; and nipple search response, 89; as the pheromone receptor system, 26–29, 30, 61

"vomeropherins," 130, 132

Whitten effect, 118–120

(Z)-7-dodecen-1-yl acetate (Z7–12:Ac), 81–82